First Lessons in Poultry Keeping
Second Year Course

by John H. Robinson

with an introduction by Jackson Chambers

This work contains material that was originally published in 1907.

This publication is within the Public Domain.

This edition is reprinted for educational purposes
and in accordance with all applicable Federal Laws.

Introduction Copyright 2018 by Jackson Chambers

The World's Largest Selection of Vintage Poultry Books

www.VintagePoultry.com

Self Reliance Books

Get more historic titles on animal and stock breeding, gardening and old fashioned skills by visiting us at:

http://selfreliancebooks.blogspot.com/

Introduction

I am pleased to present yet another title on Poultry.

The work is in the Public Domain and is re-printed here in accordance with Federal Laws.

As with all reprinted books of this age that are intended to perfectly reproduce the original edition, considerable pains and effort had to be undertaken to correct fading and sometimes outright damage to existing proofs of this title. At times, this task is quite monumental, requiring an almost total "rebuilding" of some pages from digital proofs of multiple copies. Despite this, imperfections still sometimes exist in the final proof and may detract from the visual appearance of the text.

I hope you enjoy reading this book as much as I enjoyed making it available to readers again.

Jackson Chambers

CONTENTS.

LESSON I.
INBREEDING AND LINE BREEDING — — — — — — — 5

LESSON II.
SOME COMMON PHENOMENA OF BREEDING — — — — — — 16

LESSON III.
MARKET DUCK CULTURE — — — — — — — — 24

LESSON IV.
GOOSE CULTURE — — — — — — — — — 34

LESSON V.
TURKEY GROWING — — — — — — — — — 43

LESSON VI.
POSSIBILITIES AND PROBABILITIES IN POULTRY CULTURE — — — 52

LESSON VI. — SECTION II.
SOME TYPICAL VENTURES IN POULTRY KEEPING — — — — — 61

LESSON VI. — SECTION III.
BRANCHES OF POULTRY CULTURE AND CLASSES OF POULTRY KEEPERS CONSIDERED IN THEIR RELATION TO PROSPECTIVE POULTRY KEEPERS' EXPECTATIONS OF SUCCESS — — — — — — — — — — 70

LESSON VII.
LOCATING AND LAYING OUT POULTRY PLANTS — — — — — 75

LESSON VIII.
KINDS, BREEDS, AND VARIETIES OF FOWLS — — — — — 86

LESSON IX.
STOCKING THE POULTRY PLANT — — — — — — — 96

CONTENTS.

LESSON X.
The Most Important Part of the Poultryman's Equipment - - - 102

LESSON XI.
The External Parasites of Poultry - - - - - - 106

LESSON XII.
Internal Parasites of Poultry - - - - - - 110

LESSON XIII.
External Characters of Poultry, and Their Values - - - - 114

LESSON XIV.
Poultry Nomenclature and Abbreviations - - - - - 125

LESSON XV.
Eggs and Egg Production - - - - - - - 129

LESSON XVI.
Some Elementary Moral Science for Exhibitors - - - - 133

LESSON XVII.
Business Morality in Poultry Culture - - - - - - 139

LESSON XVIII.
Winter Egg Production - - - - - - - - 146

LESSON XIX.
First Treatment of Sick Fowls - - - - - - - 150

LESSON XX.
Poultrymen's Organizations - - - - - - - 154

Index - - - - - - - - - - 158

NOTE.— First Lessons in Poultry Keeping appeared in serial form in FARM-POULTRY, in 1905. The Second Series of Lessons in Poultry Keeping, which takes up some advanced subjects, and some necessarily omitted the first year, was published serially in the same paper in 1906. The third year's course now appearing in FARM-POULTRY, describes and discusses special branches and combinations in poultry culture with the same fullness of detail and explanation as the Lessons in the earlier series.

Lessons In Poultry Keeping.

SECOND SERIES.

LESSON I.

Inbreeding and Line Breeding.

IN TAKING up these two closely related subjects, I anticipate that the treatment of them which the facts bearing upon them require will be in a measure disappointing to many readers. Outside of those who have made a somewhat thorough study of the principles of breeding, comparatively few people can be found who will come to a consideration of the subject without a prejudice against inbreeding, while among novices in poultry breeding "line breeding" is very commonly regarded as an elaborate scientific system of mating by which expert fanciers produce their finest specimens.

Both of these attitudes are wrong. The prejudice against inbreeding, though ages old, almost universal, and supported by religious and social teachings, will not stand impartial investigation. A reasonable test of inbreeding, with due regard to recognized general principles of breeding, while it will not invariably give results which confound the claims of those who oppose it, will do so often enough to discredit their arguments, and demonstrate that their position is not tenable.

Systematic line breeding, according to a prearranged "scientific" system designating the exact lines to be followed, and approximately the individuals to be used, is rarely practiced by the best breeders. Their "line breeding" plans are designed not to perpetuate certain blood lines, but to secure certain results or establish certain characteristics. If the results they seek can be best obtained within the blood lines which they are breeding they keep to those lines, or rather within certain lines; but they regard adherence to a particular line of breeding as a good thing only as long as it furnishes the best route toward the goal they seek—the best means of attaining the ends for which they are working. At any time that they see results in other lines which make it appear that blood from those lines might be introduced into their stock to advantage, they try to get such breeding specimens of those lines as seem most suitable for their purpose. They use this new blood at first experimentally to learn what results will come from the introduction of the new blood. If results are satisfactory the new blood is gradually distributed through their whole stock, and with its introduction begins a new general line of breeding to be followed only as long as it gives better results than are to be obtained by other matings.

Now while the breeder does not bind himself to follow a certain line or system, and, theoretically, is free to introduce new blood at any time, in practice he does so only at long intervals. The reasons for this will appear further on in this lesson. The two facts I wish to impress before taking up the special consideration of each of our subjects are:

    1.— That inbreeding is neither a necessary evil nor necessarily an evil.

    2.— That there is no virtue in line breeding except as it secures and fixes certain desired results.

What is Inbreeding?

Inbreeding is breeding from fowls that are near kin. If the practice is continued through a number of generations it is in-and-in breeding—if one wishes to be precise in speech, though common usage tends more and more to the simpler term.

Inbreeding in fowls extends to unions of the nearest kin — sire and daughters, dam and son, brother and sisters.

Objections to Inbreeding.

Primarily the objections to inbreeding are based, as has been stated, on social and religious arguments. As these can apply in a question like this only in so far as it can be shown that they affect physical condition we need not enter into them here. It is sufficient to consider the subject solely from the physical standpoint, and as it practically concerns the poultry breeder.

It is claimed by those who oppose inbreeding that it produces certain and rapid physical deterioration. That it leads inevitably to degeneracy and sterility. They will admit that by inbreeding some superficial, "fancy" points may be obtained in greater perfection, but claim that inbreeding cannot be carried beyond a very few generations except at the expense of vigor, stamina, size, productiveness, and that if persisted in it leads surely to the extinction of the line. In support of their position they refer to instances cited by early writers of the Darwinian school, to the conclusions of leading investigators of that period, to occasional experiences with or experiments in inbreeding poultry, and to the very evident fact that a great many stocks of pure bred fowls are deficient in stamina and "practical" qualities. The fact that many mongrel stocks on farms and elsewhere which breed indiscriminately, quickly deteriorate is also cited in this connection.

How the Objections to Inbreeding are Disposed Of.

First let us take up the points just mentioned in the order in which they have been given.

Even if the later scientists of the Darwinian school, and some of its leading exponents in these latter days, had maintained the early attitude on inbreeding, it would be pertinent to observe that few of the examples upon which they based their earlier conclusions are of more authority than the ordinary newspaper story of like occurrences, and that none (I believe) will stand such a test as would be required were they presented today as new evidence.

But, inasmuch as the leading scientists of this school, on further investigation of the subject, concluded that inbreeding did not necessarily lead to bad effects, and modified greatly their earlier views about the advantages of crossing—that is, of uniting wholly unrelated or different lines of blood, the reference to their views is chiefly interesting as showing how little the condemnation of inbreeding rests upon a careful and thorough investigation of the subject.

What is said of experiences and experiments in inbreeding merits more attention. As to the experiences: There is no doubt that there have been countless instances where inbred fowls have shown lack of stamina and general deterioration. But we have to take account also of like results where inbreeding is carefully avoided. Any impartial observation covering a great many cases will show that these results are so common where inbreeding is avoided as to suggest that the real cause or causes of them must be sought elsewhere.

Experiments in inbreeding poultry made by those who advise against it have — so far as I have seen the reports of them — invariably shown quickly and unmistakably the evil effects they anticipated; but I have never yet seen a report of such experiment which showed that it had been planned with an understanding of what was required to make a test, nor have I ever known an instance where a man who had gone on record in condemnation of inbreeding either attempted to reply to criticism of his experiment which showed where it failed, or by further experiment with conditions corrected attempted to verify his results. That this statement constitutes a severe arraignment of the advocates of what (as far as the numbers who accept it go) is the popular side of an important question, I am well aware; but as a poultryman who, after years of breeding in which inbreeding was carefully avoided, because the weight of authority on poultry culture seemed to lie that way, began to test the matter for himself, and as a result of repeated experiments arrived at the conclusion that the opposition to inbreeding had little real support either in fact or reason, I make the statement deliberately, for I question whether there is any other important matter upon which poultrymen take opposite views, where the side having the greatest numerical support — the popular side — can muster so slight a support in facts and logic.

To the general argument that the stocks of fancy fowls said to be inbred are often lacking in stamina and practical qualities as a result of inbreeding, the reply is that though so often assumed as the cause of deterioration, no one to my knowledge has ever even attempted to show it — to prove it in any case. The reasoning in this connection invariably is that if inbreeding has been practiced that is all that is necessary to account for deterioration. Everything wrong is charged to inbreeding, and no effort made to connect a trouble with any other cause, though there may be other and more probable causes obtruding themselves upon the attention of anyone making even a most cursory study of the case. The statement already made about experiences in inbreeding applies with equal force to the general argument about lack of vitality in inbred stocks. Such lack of vitality and the loss of practical qualities which often goes with it are not the peculiar and inevitable results of inbreeding, but are the common results of breeding from poorly developed specimens that are good in some superficial point prized by fanciers, and are just as likely to be found in the progeny of birds that are no kin as in the offspring of those that are near akin.

The use of flocks in which indiscriminate breeding is the rule to illustrate an argument against inbreeding furnishes no substantial support to such an argument, for it is just as appropriate to assign degeneracy in such flocks to violation of principles of breeding about which there is no dispute.

The truth is that the case against inbreeding does not rest upon trustworthy evidence, but derives most of its strength from the prejudice against it which poultry breeders almost without exception bring with them to their first efforts in breeding fowls. Before a breeder is in a position to begin to breed intelligently, and to put in practice the principles of breeding, he must divest himself of all prejudice in the matter, and consider it at least possible that inbreeding is a safe as well as a sure way of arriving at certain results. In the statement of the case for inbreeding which follows, I will try to make the reasons which justify it as clear as the facts which disprove the general argument against it will be to anyone who gives it a full and fair test. To those who will not be convinced by argument and statement of facts I can only recommend a practical test.

Inbreeding and Practical Qualities.

That "like begets like," is the fundamental principle in breeding for definite purposes. Every theory, rule, principle, law, method, and system of successful production of live stock depends more or less on this fundamental fact, and must take account of it.

An interpretation of facts which fails to consider it, a rule which excludes it, advice opposed to it, is theoretically unsound, and practically defective.

The idea that inbreeding is useful or necessary to develop "fancy" points, but detrimental and destructive in other respects involves two absurd assumptions:

      (1).—That in the progeny of a union of fowls (or animals) near akin the universal law that "like produces like," is operative in regard to one class of characteristics, but inoperative with respect to another.

      (2).—That in the progeny of a union of fowls (or animals) not near akin, the law "like produces like," operates as to the qualities in which in matings of related fowls (or animals) it is not operative.

Such contradictions in the facts bearing upon the operation of a supposed law or principle would, if they actually existed, make the practical application of that law or principle so unreliable and uncertain that it could not be made a fundamental principle.

But there is no such contradiction in the facts. It has repeatedly been shown that "like produces like" in practical or substantial, as well as in fancy or superficial qualities, both when the individuals used in the mating are near akin and when they are no kin.

This does not mean that the progeny of a mating are exact duplicates of their parents, or uniformly of an intermediate type, or that desired qualities are always reproduced. In operation the law "like produces like," is modified, as we shall see, by the fact that it includes the more remote as well as the immediate parents, but, as we shall also see, these modifications of and seeming variations from the law are strictly in accordance with it, and afford no basis whatever for assuming that the law fails to operate in regard to any characteristics or qualities.

That it often happens that fowls in which some superficial qualities have been improved or intensified by inbreeding have at the same time deteriorated in size, stamina, and productiveness, no one familiar with the phenomena of breeding poultry would deny. But — as has already been pointed out — it also often happens that size, stamina, and productiveness are diminished in fowls that are not inbred. In such cases the explanation given — and it is the correct explanation — is that lack of size, stamina, and productiveness are due to breeding from birds lacking in these points. They may also be due to external causes, but these need not be considered here.

Now if in mating unrelated fowls it is necessary to select for size, stamina, and productiveness, if we wish to reproduce those qualities in the offspring, it is also necessary to use the same care in selecting for matings of related fowls. And if by selecting for a mating of unrelated fowls, specimens having the size, stamina, and productiveness we want in their offspring, we, as a rule, get those qualities in the offspring; then if we select in the same way for a mating of related fowls, we may reasonably expect to get offspring like their parents in these essential qualities. We not only may reasonably expect to get them, but we do get them as regularly as we get results in any other kind or class of qualities. And the reason many who inbreed for fancy points, and some who make experiments in inbreeding, note a loss of "practical" qualities is that they failed, in making their matings, to provide for the retention of those qualities.

To cite the available evidence in support of the propositions I have just laid down would extend this lesson beyond reasonable limits, but I can assure the reader that every statement I have made can be amply supported, and also that if he needs or wants first hand proof he can readily obtain it by testing the matter for himself, observing that a proper test requires that the immediate parents be not deficient in any quality it is desired to have conspicuous in the progeny.

Why Inbreed?

So far I have been trying to show that inbreeding is not necessarily a cause of deterioration in practical or in any other qualities, and so to convince the reader that inbreeding is a legitimate and useful means in breeding poultry. Now I want to take the reader one step further and show him that inbreeding is not merely a method to be considered equally with the practice of constantly or periodically introducing new blood, but that it is a better method — in fact, the only method by which high excellence and uniformity can be reached and maintained.

Reference has been made to the fact that the operation of the law "like produces like," is not limited to the qualities of the immediate parents, but include also those of more remote ancestors.

According to a law, known as Dalton's law of heredity, based upon observed facts of heredity, the general rule is that an individual inherits:

One-fourth of his qualities from each parent.

One-sixteenth from each grandparent.

One sixty-fourth from each great-grandparent.

One two hundred and fifty-sixth from each great-great-grandparent.

An individual has two parents, four grandparents, eight great-grandparents, sixteen great-great-grandparents. Then

From 2 ancestors in the first preceding generation he inherits one-half of his qualities and characteristics.

From 4 ancestors in the second preceding generation he inherits one-fourth.

From 8 ancestors in the third preceding generation he inherits one-eighth.

From 16 ancestors in the fourth preceding generation he inherits one-sixteenth.

From 30 ancestors in the four preceding generations he inherits fifteen-sixteenths of his characteristics and qualities, leaving but one-sixteenth of his inheritance to come from the progres-

APPLICATION OF THE LAW OF HEREDITY.

sively increasing numbers of ancestors in all earlier generations. From each of the 64 ancestors in the fifth preceding generation an individual inherits only approximately one five-thousandth of its characteristics. Hence the influence of this and earlier generations is so slight that it may be disregarded. An undesirable feature which had not appeared in a stock for five generations is practically eliminated.

The mathematical statement of this law has never—that I know of—been applied to poultry breeding for the purpose of verification. Breeders familiar with the law observe that results in breeding seem on the average to correspond very closely with the results indicated by this law, and are satisfied to accept it.

It should also be said that a parent does not transmit alike to all offspring. There is no absolute rule—at least none that is known, and in breeding for special results expectations are based on the probability that a certain proportion of the offspring of a mating will have the qualities or characters sought.

To illustrate:—Suppose a breeder of feather legged fowls finds in his flock a pullet that is absolutely clean legged. We need not inquire here how such a pullet might be found in such a flock, and be of the same breeding as the others. It might happen. From this pullet he concludes to try to make a clean legged strain of the variety. He mates her with a male of the same breeding having the lightest leg feathering. He hopes from such a mating to get a few chickens bare legged like the hen. Most of the chickens he expects to come more or less feathered on the legs. What would be of most use to him would be a cockerel clean legged like his dam.

According to the general law stated above he may reasonably expect about one-fourth of the progeny of this hen to inherit from her, but whether they are to be bare legged or inherit other qualities is uncertain. But suppose he gets one bare legged cockerel. Next season he mates this cockerel with his dam, and may reasonably expect a good proportion of the progeny of this mating to have clean legs, for both parents and one grandparent have that feature. From this mating he may have clean legged specimens of both sexes, and mating these together he may expect clean legs to preponderate, for the parents, the grandparents, and half the great-grandparents were clean legged.

Now suppose that not seeking to make a clean legged stock the breeder carelessly, or to get some other quality of the clean legged bird which he desires to fix in his stock, or, suppose that by some chance mating of which he may not know the blood of a clean legged fowl is introduced into a line of feather legged ones. Clean legs and scantily feathered legs are bound to appear in the progeny for several generations, no matter how careful he may be in future matings, but if he avoids using them, and is careful not to introduce blood in which the same fault appeared more recently than in the line from which he is trying to eliminate it, it will take him only a few years to get it virtually stamped out. The rare chance of inheriting it from the ancestor that brought it into the line may bring a clean legged specimen occasionally for many generations, but they will not come in sufficient numbers to seriously affect results.

Let us make now the general application of the law we have examined as it affects a single quality.

The breeder of fowls has to deal with many desirable and many undesirable qualities. If he breeds systematically to definite standards for a number of years he secures a certain uniformity of desirable qualities, and may also have a similar uniformity of undesirable qualities, though from the fact that his selection with regard to undesirable qualities is a negative selection that is likely to be much greater variety in faults (as he considers them) than in merits.

He may mate together a male and female that are as nearly alike as possible in every respect. In proportion as they are bred on the same lines, and have the same ancestors or many of the same ancestors in common, they may be expected to produce chicks uniformly like themselves. But if they are entirely unrelated the chances of their reproducing their type are very much reduced, and if — as is often the case — one or both of them came from stock in which the ancestry presents a variety of different types, the results are apt to be very discouraging, for, as the law given indicates, a fowl may have thirty different ancestors, each of which may have an appreciable effect on his inheritance of qualities.

By inbreeding, by mating fowls bred on the same lines, the number of ancestors is reduced, and thus the number of different kinds and degrees of attributes which the fowl may inherit, are reduced, while its inheritance of qualities common in its ancestry, is increased and intensified. Of course this applies to faults as well as to merits. It is because inbreeding increases or fixes the faults as well as the excellencies of the line that when it is practiced by those who do not give proper attention to selection to avoid weaknesses, or whose methods of handling fowls are injurious, it may make their stock deteriorate more rapidly than if they were constantly bringing in new blood. Indeed the frequent introduction of new blood tends to a general mediocrity in the stock, without either striking excellence or marked degeneracy in any respect.

The breeder, however, is not working for mediocrity, but for excellence, and the highest excellence obtainable. To get this experience has demonstrated that inbreeding — and very close inbreeding — is necessary.

What is Line Breeding?

Line breeding may mean many different things. The phrase is used very loosely. It is common in advertisements and circulars. Breeders speak of their stock as "line bred," or line bred for so many years. So used the term conveys no definite information.

In varieties in which special matings are used to produce exhibition specimens of the different sexes, each sex is produced according to a general system of line breeding, the males and females of the different lines being of distinctly different color types. Often a breeder of such varieties, speaking of his stock as bred in line means only that his stock has been bred always from birds of the appropriate type and general line of breeding.

Again, when a breeder says he breeds in line he may mean only that his present stock contains some of the same blood as that with which he started, or as that from which he dates his line breeding. The stock may not have been bred at all systematically, but he calls it line bred because he can follow a certain line of blood back through it.

But systematic or "scientific" line breeding is something quite different. As a rule it begins in the discovery of a single bird of unusual excellence and breeding power, or prepotency. The breeder who is intelligently seeking for certain results may make many efforts to start a satisfactory line of breeding, but not until he begins to get satisfactory results does he settle down to one line. The others are merely tentative.

Having produced, or procured, and discovered through its progeny a specimen fit to become the head of a line, the breeder proceeds systematically to perpetuate this line. He studies to get the type of the opposite sex best suited to use with his phenomenal bird to reproduce its excellencies. Its finest offspring of the same sex especially are mated as far as possible to maintain in at least a few of each generation the highest possible development of the excellence reached in it. At the same time other matings are made both along the same blood lines, and with promising combinations, that in case at any time the main line, or the direct line as maintained in the finest breeding specimens in each generation should prove unsatisfactory or need reinforcement of the same line of blood, there may be abundant material from which to select.

Breeding in this way many of our best breeders continue a single line of breeding through many years. Sometimes it is a male line that is kept unbroken; sometimes a female line. Sometimes there is not direct continuity in either male or female line, but an irregular alternation according to the judgment of the breeder as to the best way to use available birds.

Rarely is the breeding according to a prearranged schedule. Results of matings are too uncertain for that. The successful matings, however, and those which produced birds which became of importance in their line are a matter of records, which constitute in a general way the pedigree chart of the stock. This, briefly, is line breeding as practiced by the most successful breeders and fanciers. They breed closely, often breeding in and in, again and again, but always intent on the points of excellence they prize most, and never maintaining a line merely for the sake of continuing it. As I stated at the outset, with the intelligent breeder a system is a means to an end, and any special system or line of matings is to be followed only as long as it appears to be the best means to gain the ends sought.

Theoretic Line Breeding.

We come now to line breeding as most beginners and some more advanced students of breeding problems want it: That is, line breeding according to a prearranged schedule, the special object of which is to avoid the evils of inbreeding.

Having read what I have said on the subject of inbreeding, the reader hardly needs to be told that personally I believe the rational way to avoid the evils alleged to result from inbreeding is by constant rigid selection, and that if this is practiced a systematic plan of producing specimens of certain degrees of consanguinity to be used at pre-ordained stages of the system is superfluous, and as this belief rests on repeated tests in my own flocks as well as a considerable volume of testimony from others, to me the practical value of work of this kind seems very small. In my mind the whole science of breeding resolves itself into the selection of the best specimens, and provided due attention is given to all essential qualities the matter of relationship may be safely disregarded. In practice, the selection of the most suitable specimens to mate together will generally keep a breeder who has a large stock of high degree of merit within his own stock without often calling for consecutive matings of very closely related birds. The rule of making strength in one sex compensate for weakness in the corresponding section in a mate of the opposite sex, if followed, keeps the breeder clear of the error of mating related birds having the same serious defect, a mating which is wrong whether birds are related or not.

However, for the information of those interested, I reproduce herewith I. K. Felch's "Breeding Chart" and an illustration of the practical application of it, which he furnished this paper several years ago. Even though one may not think it necessary to adopt such a system for the purpose of modifying the effects of inbreeding familiarity with a chart like this may be made useful in several ways, and especially in indicating a method of diagramming a record of matings, and in illustrating the variety of matings that may be made using the same lines of blood. Such a chart also illustrates admirably the point made in the discussion on inbreeding of the reduction of the number of ancestors accomplished by inbreeding.

The explanation of the chart and the system I condense from two articles on the subject by Mr. Felch. It may be noted that apart from the difference of opinion as to the effects of inbreeding, Mr. Felch's advice about breeding strongly reinforces what I have said about the necessity of care in selection, and the importance of considering a line or system of breeding as the means, not the end. Mr. Felch says:—

We know that we can take a single pair, and raise thousands in the same vigor, form, and beauty of plumage as the pair we start with; but this demands that our selections shall be of the best and most healthy specimens of their race, and that they shall be kept under the most favorable conditions.

Our chart shows how a pair and its progeny can be bred, without resorting to breeding that would be termed incestuous. The art of this consists in being able at all times to produce a flock that is one-half the blood of the original pair.

By a careful examination of the chart it will be seen that all groups to the right of the center have a preponderance of the blood of the male used in the first mating, and all groups to the left of the center have a preponderance of the blood of the female, the figures showing just what the proportion of blood is, while the center groups are each and all just one-half the blood of each of the original parents.

Experience has taught us that to breed for three generations the same blood is disastrous. Were we to mate male and female from group 3, and to repeat the mating in the progeny for three generations we would produce sterility and lack of vigor. But our groups 3, 7, 11, and 16 are all one-half of the blood of each ancestor, and sound, vigorous, and productive, because of the method of their production, and just as strong as group 3, which was the direct issue of the original mating.

Throughout we persist in an unbroken line of males, as demonstrated in the black lines of the chart. We establish a line of breeding that will be wonderful in its like producing like quality. In the first product, group 3, there may not be a single male to present the type of No. 2, nor may there be a pullet in the type of No. 1, but by breeding back we secure the original types of both, by which when we make the second group of one-half bloods we find both males and females to our liking.

12　　LESSONS IN POULTRY KEEPING — SECOND SERIES.

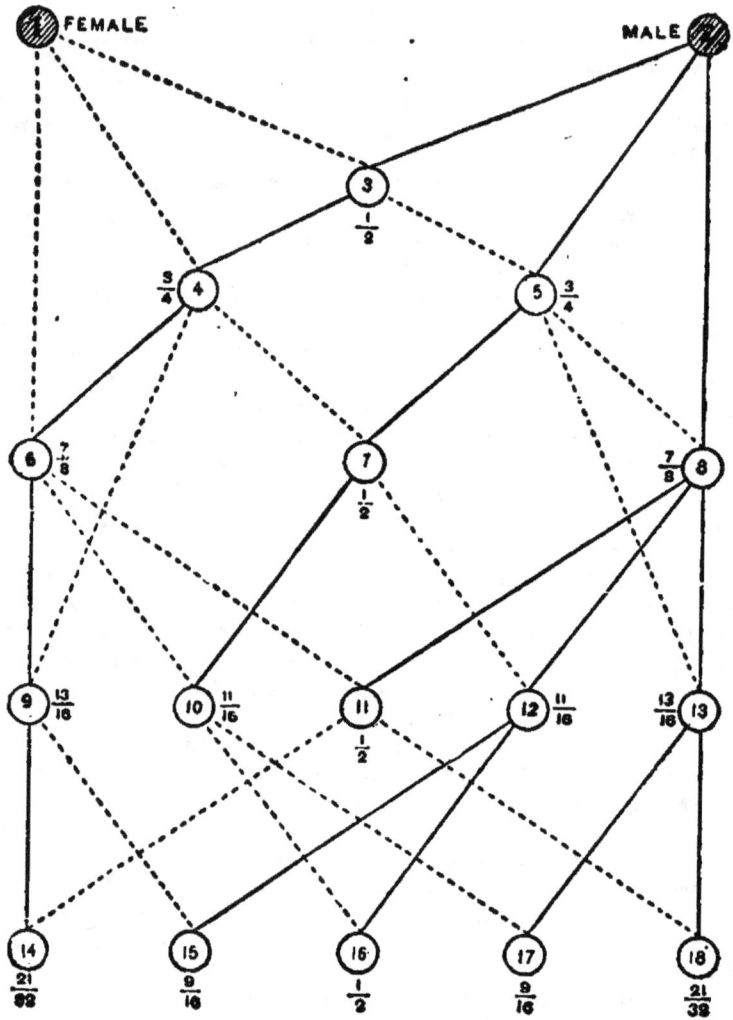

Felch Breeding Chart.

Or it may be that in group 3 we have discovered a type we admire above that of the original pair. If so, such should be our standard, and in subsequent matings we should take care to select only birds of that type, until all our matings result in that uniform type, size, and color.

To make the chart clear we say that each dotted line represents the female as having been selected from the upper group to which it leads, while the solid line shows the male as having been taken from the upper group indicated. Each circle represents the progeny of a male from the upper group to which the solid line from it leads, mated with a female from the upper group to which the dotted line from it leads.

Female No. 1 mated with male No. 2 produced group No. 3, which is one-half the blood of sire and dam.

Females from group No. 3, mated back to their own sire No. 2, have produced group No. 5, which is ¾ of the blood of the sire, No. 2, and ¼ the blood of the dam, No. 1.

A male from group No. 3, mated back to his own dam, No. 1, produces group No. 4, which is ¾ of the blood of the dam, No. 1, and ¼ the blood of the sire, No. 2.

Again we select a cockerel from group No. 5 and a pullet from group 4, or vice-versa, which will produce group 7, which is mathematically half the blood of each of the original pair, No. 1 and No. 2. This is a second step towards producing a new strain.

APPLICATION OF BREEDING CHART. 13

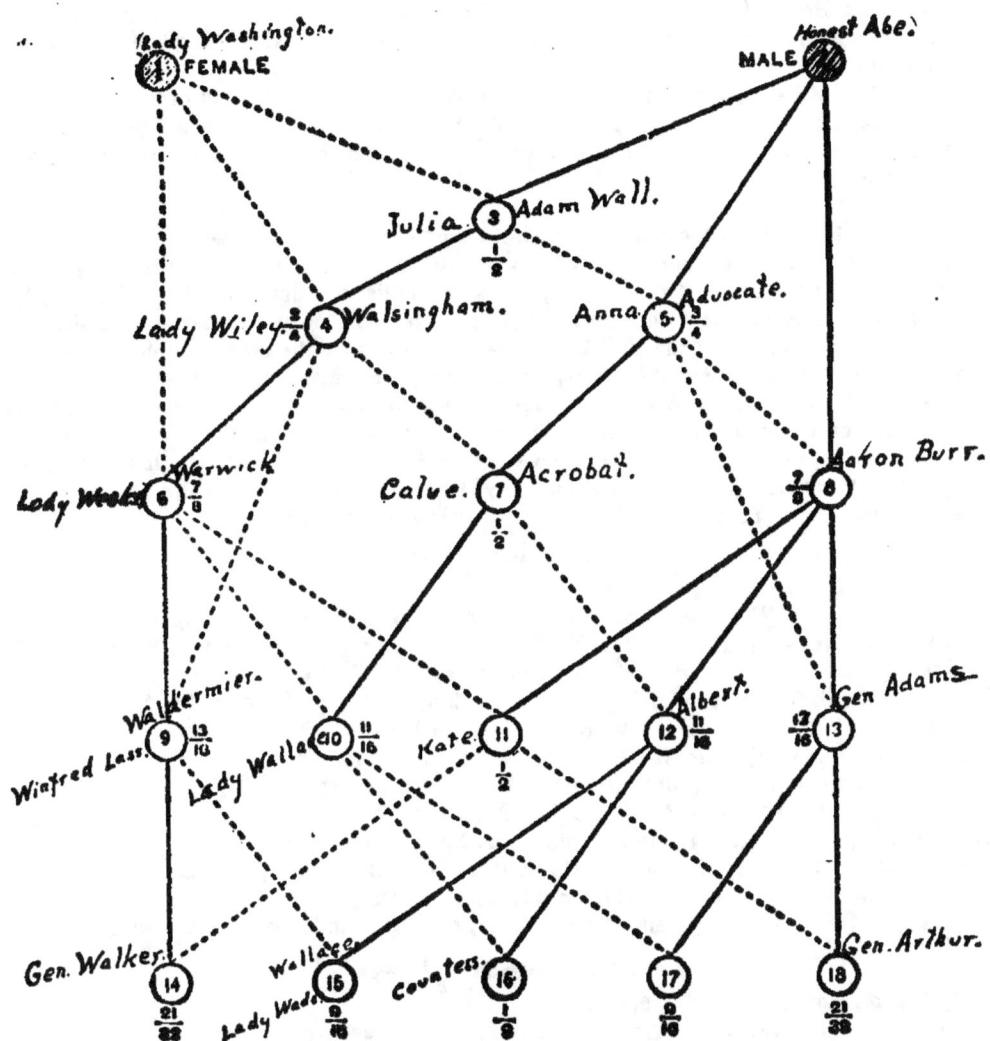

Felch Breeding Chart Applied to Line Described in Text.

Females from No. 8 mated back to the original male, No. 2, produce group 8, that are ⅞ the blood of No. 2, and a cockerel from No. 4, mated back to the original dam, No. 1, produces group No. 6, that is ⅞ the blood of the original dam, and only ⅛ the blood of the original sire.

Again we select a male from No. 8 and females from No. 6, and for a third time produce chicks (in group No. 11) that are half the blood of the original pair. This is the 3d step, and the 9th mating in securing complete breeding of our new strain. In all this we have not broken the line of sires, for every one has come from a group in which the preponderance of blood was that of the original sire. Nos. 2, 8, 13, and 18 are virtually the blood of No. 2.

We have reached a point where we would establish a male line whose blood is virtually that of our original dam, and we now select from No. 6 a male which we mate with a female from No. 4 and produce group 9, which is 13-16 the blood of the original dam No. 1, and 3-16 the blood of the original sire.

Again we select a male from No. 9 and a female of the new strain, No. 1, and produce group 14, which becomes 21-32 of the blood of the original dam, thus preserving her strain of blood.

A male from No. 13, which is 13-16 the blood of the original sire No. 2, mated to females from No. 10, which are 5-16 the blood of the original sire, No. 2, gives us group 17, which is 9-16 the blood of said sire.

While in No. 16 we have the new strain and in No. 18 the strain of our original sire, No. 2, we have three distinct strains, and by and with this systematic use we can go on breeding for all time to come. Remember that each dotted line is a female selection and each solid line the male selection.

To make all this plain to the novice, I shall, in showing by the pedigrees of individual fowls how to follow my diagram, number both males and females by the number of the group as seen in the chart, they being of that group; while in the list of names will be added = $\frac{1}{2}$, $\frac{3}{4}$, $\frac{7}{8}$, etc., as the case may be, the fraction showing the extent of the preponderance of the blood of the strain, the balance of the unit being the percentage of the strain mated: — all names commencing with A being of Honest Abe, No. 2 strain; all names commencing with W having in excess the blood of Lady Washington and her strain; other names representing one-half the blood of each of these two strains. This middle ground or reservoir of blood, we create of our two strains, that we may both preserve line breeding as applied to the individuals, and when we see fit to break the individual line, may still maintain the line breeding of the strains.

We now proceed to name the individuals and create their pedigrees.

The Male List.

Honest Abe, No. 2, our original sire.
Adam Wall, No. 3, ($\frac{1}{2}$); sire, Honest Abe, No. 2; dam, Lady Washington, No. 1.
Walsingham, No. 4, ($\frac{3}{4}$); sire, Adam Wall, No. 3; dam, Lady Washington, No. 1.
Advocate, No. 5, ($\frac{3}{4}$); sire, Honest Abe, No. 2; dam, Julia, No. 3.
Warwick, No. 6, ($\frac{7}{8}$); sire, Walsingham, No. 4; dam, Lady Washington, No. 1.
Acrobat, No. 7, ($\frac{1}{2}$); sire, Advocate, No. 5; dam, Lady Wiley, No. 4.
Aaron Burr, No. 8, ($\frac{7}{8}$); sire, Honest Abe, No. 2; dam, Anna, No. 5.
Waldermier, No. 9, (13-16); sire, Warwick, No. 6; dam, Lady Wiley, No. 4.
Albert, No. 12, (11-16); sire, Aaron Burr, No. 8; dam, Calve, No. 7.
Gen. Adams, No. 13, (13-16); sire, Aaron Burr, No. 8; dam, Anna, No. 5.
Gen. Walker, No. 14, (21-32); sire, Waldermier, No. 9; dam, Kate, No. 11.
Wallace, No. 15, (9-16); sire, Albert, No. 12; dam, Winfred Lass, No. 9.
Gen. Arthur, No. 18, (21-32); sire, Gen. Adams, No. 12; dam, Kate's sister, No. 11.

The Female List.

Lady Washington, No. 1, original dam.
Julia, No. 3, ($\frac{1}{2}$); sire, Honest Abe, No. 2; dam, Lady Washington, No. 1.
Anna, No. 5, ($\frac{3}{4}$); sire, Honest Abe, No. 2; dam, Julia, No. 3.
Lady Wiley and sisters, No. 5, ($\frac{3}{4}$); sire, Adam Wall, No. 3; dam, Lady Washington, No. 1.
Lady Weeks, No. 6, ($\frac{7}{8}$); sire, Walsingham, No. 4; dam, Lady Washington, No. 1.
Calve, No. 7, ($\frac{1}{2}$); sire, Advocate, No. 5; dam, Lady Wiley, No. 4.
Winfred Lass, No. 9, (13-16); sire, Warwick, No. 6; dam, Lady Wiley, No. 4.
Lady Wallace, No. 10, (11-16); sire, Acrobat, No. 7; dam, Lady Weeks, No. 6.
Kate, No. 11, ($\frac{1}{2}$); sire, Aaron Burr, No. 8; dam, Lady Weeks, No. 6.
Lady Wade, No. 15, (9-16); sire, Albert, No. 12; dam, Winfred Lass, No. 9.
Countess, No. 16, ($\frac{1}{2}$); sire, Albert, No. 12; dam, Lady Wallace, No. 10.

It now becomes easy to mate up our pens.

Pen No. 1 — sire, Honest Abe, No. 2; dam, Lady Washington, No. 1, — produces group No. 3, half-bloods.

Pen No. 2 — sire, Honest Abe, No. 2; dams, Julia, No. 3, and sisters, by Honest Abe, No. 2, produces group No. 5, three-fourths blood of Honest Abe, No. 2.

Pen No. 3 — sire, Honest Abe, No. 2; dams, Anna, No. 5, and sisters, by Honest Abe, No. 2; 2d dam, Julia, No. 3, by Honest Abe, No. 2; 3d dam, Lady Washington, No. 1, — produces group No. 8, seven-eighths blood of Honest Abe, No. 2.

BREEDING LINE OF SIRES.

Pen No. 4 — sire, Adam Wall, No. 3, by Honest Abe, No. 2, out of Lady Washington, No. 1; dam, Lady Washington, No. 1 — produces group No. 4, three-fourths blood of Lady Washington, No. 1.

Pen No. 5 — sire, Walsingham, No. 4, by Adam Wall, No. 3, out of Lady Washington, No. 1; dam, Lady Washington, No. 1, — produces group No. 6, seven-eighths blood of Lady Washington, No. 1.

Pen No. 6 — sire, Advocate, No. 5, by Honest Abe, No. 2, out of Julia, No. 3, by Honest Abe, No. 2, out of Lady Washington, No. 1; dams, Lady Wiley, No. 4, and sisters, by Adam Wall, No. 3; 2d dam, Lady Washington, No. 1; produces group No. 7, one-half the blood of each strain.

Pen No. 7 — sire, Aaron Burr, No. 8, by Honest Abe, No. 1, out of Anna, No. 5; dams, Lady Weeks, No. 6, and sisters, by Walsingham, No. 4; 2d dam, Lady Washington, No. 1, produces group No. 11, one-half the blood of each strain.

Thus do I name pedigree and mate up to and including the fourth generation, leaving the reader to mate the remainder of the chart.

Up to this point it is line bred both individually and as to the strains, and so is it line bred on the Honest Abe strain both as individuals to the end of groups 17 and 18.

But we now break the female line by producing a male of her line as her heir, and mating him, Warwick, No. 6, to Lady Wiley, No. 4, to produce Waldermier, No. 9, and his sisters, Winfred Lasses, continuing the male line by mating Waldermier, No. 9, to Kate, No. 11, thus producing Gen. Walker, No. 14, and his sisters as seen in group 14. In all this we have preserved the preponderance of the blood of Lady Washington strain.

Now the whole strength of this line of breeding stands on the breeder being able to produce a group of half-bloods from within his own flocks.

To make the strain line of breeding more plain: — Suppose we follow the rule many do of buying a foreign pullet to produce a group of half-bloods, and her set of pullets are bred back to the sire, or a male of his strain, these three-fourths blood pullets again bred back to a male of the sire's strain. This would be breeding in line to one strain, which is just half the force of our chart. These three males being different males of one strain, this would be only *strain breeding, not line breeding with an unbroken line of sires.* If our selections are carefully made, ever keeping in mind the types we start with, or the cherished types of our strain, then this breeding in line to one's strain may prove as forceful in results as adhering to a line of sires — from sire to son.

LESSON II.

Some Common Phenomena of Breeding.

IN the last lesson we discussed the subject of heredity in its relation to systematic methods of breeding to special standards. In this lesson we are to consider various phenomena of heredity as they practically concern the breeder in his work.

Since the selling and buying of eggs for hatching purposes has become so general, the old question, "Which is the mother of the chick — the hen that laid the egg, or the hen that hatched it?" has become of less interest than the question,— Which is the breeder of the chick, the man who produced and mated the stock from which it came, or the man who hatched and reared it?

Without attempting to make a final answer to that question, let me say that the production of fine fowls requires a combination of skill in mating and skill in growing. To produce the finest stock we must have first of all parent stock of fine quality, properly mated, but the product will not equal or even approach the excellence of its ancestors unless it is properly handled as it grows. It is important that the reader should keep this fact constantly in mind both when applying principles of breeding, and when looking for confirmation of those principles in the results of his matings, for if the conditions under which a fowl is grown are unsatisfactory the result may be a bird very different in shape, color of plumage, vigor, etc., from what it might have been under conditions providing for the full development of its possibilities; and conditions adverse to full natural development are apparently not limited in their effects to failure to fully develop the qualities directly transmitted from the parent stock; but there is some reason to suppose that features eliminated from a stock by careful selection sometimes reappear in ill nourished, ill cared for chickens, when they would not have developed under better conditions. Such a point as this is of course difficult to demonstrate, for we cannot show positively in what way any given individual fowl might have developed differently under different conditions. But comparison of chicks from the same breeding stock, hatched and reared by different parties, often shows them so different that it is hard to believe they could have been produced from the same parents.

Hence, for a proper appreciation of the laws of heredity, we must discriminate between results as found in well developed specimens and results as found in ill developed specimens.

Prepotency in Breeding.

As a rule, pure bred fowls transmit to their progeny a much greater proportion of their individual excellence than mongrel fowls, and pure bred fowls that have been carefully bred in line transmit their qualities to their offspring more surely than those that have been bred in a haphazard way. This is because of the cumulative breeding back of them, and is easily explained by the general law of heredity given in the last lesson.

WHAT PREPOTENCY IS.

The fancier and careful breeder, however, are not working merely for fair to good average results in the transmission of breed or variety characteristics. Their constant endeavor is to bring the average of the stock as near as possible to the best prevailing types, and while working zealously toward this end they are still seeking for new, improved, and advanced types.

The law of heredity we have discussed deals with the problems of heredity only in a general way. In its general application it seeks to secure uniformity by having an ancestry as solidly as possible in representation of the qualities sought. As far as is known there is no way of telling beforehand what qualities will appear in any mating or generation, or in what combinations or proportions. While, as has been said, it is a general fact that good breeding tells plainly in the product of a mating, it is also true that it does not show an equality of results. Some fowls have special power to mark their progeny like themselves, either in general appearance or in one or more important characteristics. This power is not dependent upon quality or stamina. It is found in fowls of all degrees of quality and vigor, but as it is prized only when found in those of some special merit it is apt to be passed without recognition in others.

A fowl which has this power to "mark" its offspring like itself is said to be *prepotent*. The combination of quality and prepotency in the same individual, while comparatively rare, occurs often enough to have a very conspicuous influence on the development of our stocks of thoroughbred fowls. A prepotent bird may transmit only its other qualities, (or some of them), to its progeny, or it may transmit its prepotency also, giving a line of stock remarkable for great excellence and the power of transmitting it.

This prepotency is not the same thing as the cumulative breeding power which is gained by careful breeding, though the two may so unite that it would be practically impossible to distinguish between them. The difference between them is easily seen when a fowl is found that is notably superior to its ancestors, and marks this superiority on its progeny to a much larger degree than our law of heredity indicates would be the average influence of a parent. It may also be seen in cases where an objectionable feature persists in a stock to a much greater extent than is usual. Though in most cases objectionable features disappear at about the rate the law indicates, it happens quite often that when a fowl with some specially good qualities and one or more pronounced faults is bred from in the expectation of breeding the good quality into a stock, and the bad quality out of it as quickly as possible, the fowl proves to be prepotent as to the bad quality as well as the good, or, perhaps, prepotent as to the bad quality and not as to the good, and the effort has to be abandoned. The prepotent fowl can be known only through the results obtained by breeding it. Some fowls seem to be prepotent however mated, others are prepotent in some matings and not in others. The reasons for this cannot be explained, but the breeder can ascertain the facts, and frequently can learn them in time to make good practical use of them. It is generally considered the best policy to continue a mating that has given especially good results as long as the fowls in it are fit for breeding. When a fowl of good quality proves an unsatisfactory stock getter in one mating, a breeder generally tries it mated differently the next season, and every season as long as fit for breeding, until he gets such results as the quality of the bird leads him to expect it might give if properly mated. In all work of this kind, the greater one's experience in breeding, and the more thorough his knowledge of the stock he is working with, the oftener he is likely to make matings that give satisfaction, though none know better than the breeders of greatest experience and skill how often their most careful matings produce stock not up to their expectations.

The Ways of Heredity.

There is quite a prevalent opinion among breeders that the female marks her progeny especially with size, shape, and practical qualities, while the male gives them color, comb, and superficial characteristics. How this opinion arose, we do not know. That it is erroneous, a little experience in crossing breeds of fowls, or observations on human families of our acquaintance, will quickly convince most people. The latter test is easier to make, for anyone can at any time make it mentally.

Students of the features of heredity recognize several very common phenomena:

> The male offspring resemble their sire, and the female offspring their dam.
>
> The male offspring resemble their dam, and the female offspring their sire.
>
> The offspring, both male and female, or either, resemble a grandparent more than either parent.
>
> The offspring may, in any point considered, inherit from both sire and dam, their quality being, in some degree, intermediate between the parents, or, in case of a quality in which the parents are alike, being in excess of either.

These are the most common and conspicuous features of heredity. In the breeding of pure bred fowls, so like in many respects, it would be difficult to trace the influences of individual parents and grandparents clearly enough to distinguish such phenomena in ordinary cases, but they may be seen occasionally when birds of marked prepotency are used for breeding, and they are traced with ease in many matings of crossbred fowls, especially in points of strong resemblance or great difference.

I think the reader is now ready to believe that the phenomena of heredity are very complex. The different features of heredity just mentioned do not occur independently and separately. All of them are sometimes observed, and in many degrees, in the progeny of a single mating, and all may be accounted for by the general law of heredity given in the preceding lesson, when considered in connection with a few other breeding principles, such as prepotency, which we have just discussed, and reversion or atavism, which will be discussed a little further on.

The point I wish to introduce here is that the law of heredity or inheritance is responsible for variations from established or favorite types as well as for the perpetuation of those types. It is often assumed that in the production of stock there are two warring tendencies— the tendency for like to beget like, and a tendency to variation, that is, to the production of features differing from those found in the parents. These, supposedly different and antagonistic tendencies are for the most part simply different manifestations of the same general law of heredity, though there is no doubt that many variations in the way of special development due to especially favorable conditions are at least in part transmitted to offspring.

Let us see how variations, not due to external conditions, arise:

All that the fowl is as it begins life as a chick it is by inheritance. In every part, feature, characteristic, quality, and possibility it is like some ancestor. We found in studying the question of inbreeding that the fowl inherited an appreciable part of its qualities from some thirty nearest ancestors, and that the chances of a characteristic not found in any of these being produced were very remote. Still there is a chance that a feature long absent in a stock, though common in its remoter ancestors, may reappear in some of them at any time; and I have heard of a few instances in which it appeared that a characteristic of some very remote ancestor had reappeared in a very large proportion of the stock produced in a season. This reappearance of features supposed to have been completely eliminated is what is called *reversion*. It is rare in stock that is carefully bred and new blood introduced with caution, but is quite common when birds of similar type but very different breeding are mated, or when radically different types of fowls are crossed. The beginner who in his eagerness to avoid inbreeding buys males and females from different breeders and puts them together without knowing anything of the breeding tendencies of either line of stock, is very apt to get some chicks that "take back" to distant ancestors in which qualities not now wanted were conspicuous; and he generally concludes that some of the stock he got was not "pure." While it is most common to have but a single feature reappear, once in a while one bird or a few birds are produced like a remote ancestor in many points.

Now suppose that in place of simple reversion to a single ancestral characteristic which one familiar with the stock or with the history of the production of a variety or breed will easily recognize as such, a combination of features of two different ancestors takes place, or a mingling of an old with a modern feature. In such a case it may be possible to analyze the new character or type, but it may be impossible to do so, and in that case we have a variation which we cannot explain. It is perhaps not unnatural to suppose that what we cannot explain is due to causes we cannot appreciate or understand, but in this case I think we can see something of the general method of variation even where we cannot trace every variation to its sources.

We can control variation in just the same way, and to much the same degree as we can heredity, of which variation is one manifestation. That is, strictly speaking, we cannot control it at all, but relying on the general law— on average results—we generally secure likeness in the features we prize most, and avoid variations except in conspicuous characters. Yet in many cases, having once found that a fowl, or the fowls of a certain line are strongly prepotent in any direction, either for impressing high excellence according to established standards or for the development of new and better types that knowledge may be made of great practical and commercial value. The ability to discover and to use prepotent fowls and types that are better or more attractive than those prevailing is one of the best personal assets a breeder of fowls can have. It depends very much upon a knowledge of the phenomena and the principles of breeding. Most of our successful breeders have a pretty thorough practical understanding of the subject, though few can be found who can give a good systematic presentation of what they know about breeding, and nearly all cling to some ideas about breeding which I have little doubt they would quickly discard if once they put all their knowledge and thoughts about breeding together in such a way that the inconsistencies in them would be clear.

To the novice in breeding the first use of a knowledge of the principles and phenomena of heredity should be to give him a more correct view of the actual status of so-called pure bred stock, and through this to make him slower to condemn breeders for faults in stock bought of them. When a novice buys breeding stock it is sometimes subjected to several conditions not favorable to best breeding results. The change of climate and environment may or may not be advantageous. The effects of such changes are different with different individual fowls. The same change may be beneficial to one, indifferent to another, detrimental to another; and these results may be directly reflected in their progeny. With such small lots of fowls as are usually purchased for breeding, it may happen that all are affected alike, and if the change is in any degree detrimental the breeding results may be disappointing. The tendency in such cases is to go back to obsolete types and characters, and quite a variety of these may reappear at the same time. Sometimes these results occur at first, but after the fowls become accustomed to the change they breed right. For this reason it is best not to condemn good fowls on the first season's breeding under new conditions.

Very often the care and feeding the novice gives his breeding stock are not favorable to good breeding condition and results. Add to this the fact that many novices practice buying from different breeders to avoid inbreeding, and mate the birds without any knowledge of them, and we have a combination of causes for reversion to obsolete types and characters, and it is not difficult to account for the skepticism with which after a season or two breeding this way, many a novice regards the statements of older breeders as to the proportion of birds of good breeding quality produced from their matings.

The Mendel Law.

For several years past breeders of live stock have been showing more and more interest in some experiments made years ago by a German scientist named Mendel, which have only recently began to get the attention they deserved, but which now seem in a fair way to be considered of more importance than is actually warranted. The "Mendel law" is very glibly discussed and applied by not a few people who have not grasped the meaning of that law at all, and are equally uninformed about many of the cases in which they attempt to apply it.

Two years ago I gave in FARM-POULTRY a statement of Mendel's law, and some observations on its application to poultry breeding and practical value to poultrymen. The number in which this appeared was almost immediately out of print. So many calls for the article have been made that as it is right in line with the subject of this lesson, and as it seems advisable to have it in permanent form, I reprint here that part of the article relating exclusively to the Mendel law.

As preliminary to a correct understanding of the scope of the investigations of Mendel, and of the possible application of his law in poultry breeding, it is necessary to understand that the greater number of Mendel's experiments in crossing were made with peas, and that his observations were in most cases confined either to noting the results with respect to reproduction of a particular character from a union of specimens differing in that character. Thus, he would cross a tall and a low variety of peas, and observe and classify resulting plants according to height. He would cross a variety having the seed round and smooth when ripe with one having wrinkled seed, and note the character of the seed from the resulting plants. And so on through a variety of minor differences. In a few cases he made comparisons for two pairs of characters, and found that the mathematical proportion discovered for the single pair was still maintained, but there was nothing in his work at all approaching in complexity the task of the poultry breeder who might undertake to make an application of Mendel's law to such number and variety of characters as we have in fowls.

Anything like a general application of Mendel's law to the phenomena of poultry breeding is at present out of the question. Indeed, though a mathematician might find mathematical expression for the application of the law to many varying characters, it would be practically impossible to separate the offspring of a mating made with such an end in view, and to identify the fowls equivalent to the factors in the mathematical statement. For the present — and perhaps for all time — poultrymen must work with Mendel's law as he worked with it, applying it to but one or two characters at a time.

However it may seem to those not versed in the histories of breeds and varieties, to experienced breeders this limitation of the application of the law will not appear to diminish its practical value. The experienced breeder knows — none better — how all but impossible it is to make rapid advance in the development of more than one important feature at a time. What the Mendel law gives promise of being able to do for the breeder of poultry is to help him to make more rapid and more certain progress point by point. As Bateson puts it, Mendel's "work relates to the course of heredity in cases where *definite varieties differing from each other in some one definite character are crossed together.* * * * It was found that in each case the offspring of the cross exhibited the character of one of the parents in almost undiminished intensity, and intermediates which could not be at once referred to one or other of the parental forms were not found.

"In the case of each pair of characters there is thus one which in the first cross prevails to the exclusion of the other. This prevailing character Mendel calls the dominant character, the other being the recessive character.

"That the existence of such 'dominant' and 'recessive' characters is a frequent phenomenon in cross breeding, is well known to all who have attended to these subjects.

"By letting the crossbreds fertilize themselves Mendel next raised another generation. In this generation were individuals which showed the dominant character, but also individuals which presented the recessive character. Such a fact also was known in a good many instances. But Mendel discovered that in this generation the numerical proportion of dominants to recessives is on an average of cases approximately constant, being in fact as three to one. With very considerable regularity these numbers were approached in the case of each of his pairs of characters.

"There are thus in the first generation raised from the crossbreds seventy-five per cent dominants and twenty-five per cent recessives.

"These plants were again self-fertilized, and the offspring of each plant separately sown. It next appeared that the offspring of the recessives remained pure recessive, and in subsequent generations never produced the dominant again.

"But when the seeds obtained by self-fertilizing the dominants were examined and sown it was found that the dominants were not all alike, but consisted of two classes: (1) those which gave rise to pure dominants; and (2) others which gave a mixed offspring, composed partly of recessives, partly of dominants. Here also it was found that the average numerical proportions were constant, those with pure dominant offspring being to those with mixed offspring as one to two. Hence it is seen that the seventy-five per cent dominants are not really of similar constitution, but consist of twenty-five which are pure dominants, and fifty which are

EXAMPLE OF THE MENDEL LAW.

really crossbreds, though like the crossbreds raised by crossing the two original varieties, they only exhibit the dominant character.

"To resume, then, it was found that by self-fertilizing the original crossbreds the same proportion was always approached, namely, 25 dominants, 50 crossbreds, 25 recessives. * * *

"Like the pure recessives, the pure dominants are thenceforth pure, and only give rise to dominants in all succeeding generations studied.

"On the contrary, the 50 crossbreds, as stated above, have mixed offspring. But these offspring, again, in their numerical proportions, follow the same law, namely, that there are three dominants to one recessive. The recessives are pure like those of the last generation, but the dominants can by further self-fertilization and examination or cultivation of the seeds produced, be again shown to be made up of pure dominants and crossbreds in the same proportion of one dominant to two crossbreds."

In illustrating the application of Mendel's law, Prof. T. H. Morgan, in an article on "The Determination of Sex," in the *Popular Science Monthly*, for Dec., 1903, makes this example: "If a white mouse is crossed with a wild gray mouse all the offspring of this cross will be gray like the wild mouse. The gray color of the gray mouse is said to be *dominant*, and the white color (inherited from the other parent) does not appear, but is supposed to be present in a sort of latent condition. It is said to be *recessive*. If now these primary hybrid mice are interbred, some of their young will be white, and the rest gray in the proportion of one to three. If these white mice, when they become grown, are interbred, their offspring will always be white as well as all their subsequent descendants. Some of the gray mice will also breed true, but the rest that are gray hybrids will, if interbred, give rise to some white and some gray in the proportion again of one to three."

We do not understand that in this illustration Prof. Morgan assumes to state facts about the crossing of white and gray mice. We take it that the case is an assumed one, except perhaps as to the statement that the progeny of the first cross would be all gray.

Now it is a question which some poultrymen may be able to answer partly from past experience in breeding, whether Mendel's law will apply to any characters in poultry, and if so, to how many and to what characters. The writer has made a good many crosses, observing results chiefly with reference to the laying and table qualities of the stock produced, but in connection with these things has taken casual notice of other points which might be supposed to come under the operation of the Mendel law — if that applies. (Bateson takes pains to emphasize and to reiterate that the law does not appear to be of universal application). We have observed some results, for instance, with regard to color or some other character which suggested that Mendel's law might apply in some cases — matings of certain individuals — and might not apply in similar cases in which different individuals were used. For instance, we once made a cross of White Leghorn on Light Brahma in which all the progeny were white, the color of the Leghorn sire. We have had other crosses in which the colors of the females of the male's line were approximately reproduced in his female offspring, while the male offspring resembled the color of males of the variety of their dams. Again we have had both males and females, without exception, of color type intermediate between the variety colors of sire and dam, and also intermediate in size, in shape, in size of comb, in size of tail, etc.

But these were not crosses of varieties. They were crosses of breeds. To reproduce Mendel's experiments with reference to color of fowls we must take two varieties differing only in color, as, say, the Black and the White Wyandotte, Leghorn, Hamburg, Cochin, Langshan, or Minorca. While we cannot at present refer to a record of such a cross, our impression is that cases which would seem to confirm the Mendel law are extremely rare. The variety of results we have seen in color, considered in connection with the limited range of Mendel's experiments, suggests that though he simplified his statement by the introduction of the terms "dominant" and "recessive," the introduction of those new terms made the relations of his observations to certain older principles obscure. Bateson calls attention to the use of the terms "dominant" and "recessive" as a clever avoidance of "the complications involved by the use of the expression 'prepotent,'" but it seems to us that by ignoring "prepotence" the facts are placed on a false basis, for it is not possible that the law may be found to apply to the phenomena of prepotency rather than to heredity in general. Bateson approaches, but does not reach this conclusion when in the more technical discussion of the subject to which he devotes the latter half of the paper from which we have quoted, he mentions a number of cases which, like those in crossing poultry which we have referred to, plainly do not accord with Mendel's results, and follows with this conclusion: "Dominance, as we have seen, is merely a phenomenon incidental to specific cases, between which no other common property has yet been proved. In the phenomena of blended inheritance we clearly have no dominance. In the cases of alternative inheritance studied by Galton and Pearson there is evidently no universal dominance."

Now if the principles discovered by Mendel are not of wider application than to the crossing of such definite varieties differing in one or a few definite characters, they would be of little importance to poultrymen generally, for very few poultrymen are practically interested in the development of types from crosses of varieties. The making of new breeds and varieties of poultry is generally accomplished by mingling several quite distinct breeds often differing from each other in nearly all sections. Each breed or variety used is used for a definite purpose—to introduce some special character or feature of its own, or to make in combination with another some new character intermediate between the two.

It has long been a common saying among poultrymen that in a first cross we get a certain uniformity, but that in breeding together the progeny of this first cross we get a great variety of results. Outside of the few breedmakers, those who make crosses have generally given up trying to get uniformity out of the progeny of a cross, and if they like a certain cross content themselves with renewing it as often as necessary. It cannot yet be said that the Mendel law suggests general rules for, or plans which may be universally applied in bringing orderly development out of the seeming confusion produced by the breeding together of crossbred fowls. Nor can it be said that what it does suggest as to how a mating, which produces a desired character as a dominant character, may be used and followed up to best advantage is at all new to poultrymen. It does, however, suggest the introduction of a new method in breeding which no doubt has sometimes been used accidentally, but which, so far as we are aware, no breeder has ever recognized or advised. And if the Mendelian principles are demonstrated to apply to the phenomena of breeding generally, and not merely to the crossing of distinct varieties, their influence on breeding operations through the introduction of this new method cannot fail to be of great practical value.

To explain what we mean:—There does not seem to be any good reason why we should not consider what Mendel called "dominant" characters as prepotent characters. In their breeding operations poultrymen have been accustomed to attach importance only to those individuals possessing the character they sought to fix, and have found that some birds will reproduce that character, and some will not. By continued breeding and careful selection they finally eliminate all the specimens that will not reproduce the desired character acceptably, but in any case in which the principles discovered by Mendel operate the breeder who works for positive results begins his work with, say three-fourths of all the stock from a certain mating showing a decided character, while only an unknown third of that three-fourths will reproduce the character with certainty. On the other hand—if the principle of Mendel applies—all of the stock which has this character different will reproduce the different character in its progeny.

Certainty in results is secured by working not with the individuals having the dominant character, but by using those having the recessive character.

The poultry breeder generally discards those specimens which differ in any desired quality from their parents – if he knows them. Sometimes he uses a "chance" bird and finds him a very strong breeder. May it not be because in respect to a certain quality or qualities he is what Mendel calls a "recessive?"

And is it not worth while for poultrymen to study their matings in the light of Mendel's discoveries and see whether the "science of breeding" cannot be made more accurate and satisfactory by seeking out and using the specimens that retain a desired quality when most of their kin in the same generation lose it rather than by working, especially with the specimens which come from the mating most fruitful in producing specimens with any given desired characteristic?

To illustrate by an example:—

Suppose a breeder of fowls of a five toed variety secures two males of that variety having only four toes on each foot, and with these males as a starting point proposes to change his stock from five toed to four toed.

Now if Mendel's law is a law of prepotency he might get opposite results from the matings of these two males with one hen each. He might have the progeny of male No. 1 all four toed like the male, and the progeny of male No. 2 all five toed like the female. If then on inbreeding individuals from each mating he found the variations as to toes approximately according to Mendel's results, this would be the situation: From the progeny of male No. 1 he would have seventy-five per cent of four toed chicks, and only twenty-five per cent of five toed chicks. From the progeny of male No. 2 he would have seventy-five per cent of five toed chicks, and only twenty-five per cent of four toed chicks, but,—the four toed chicks from male No. 2 would be the ones of most service to him, because he would know that they would reproduce themselves with certainty, while only an unknown in every three of the four toed chicks of the progeny of male No. 1 would be pure bred as to the number of toes required.

Thus it will be seen that if it could be established that the Mendel law applied to some things in poultry breeding, a breeder acquainted with this law who should produce from a mating, or discover that a chance mating had produced a lot of chicks divided with respect to any particular character, as the offspring of Mendel's crosses were divided, might reasonably assume that he had found an instance where the law applied, and proceed to make his matings accordingly.

We can see, too, how the discovery that this principle applied in any particular case might be of great importance in the preservation of established characters. One of the greatest difficulties in breeding poultry is to hold points gained while making changes in other points. The knowledge that certain individuals, or the entire progeny of a certain mating, or line of matings, was especially strong in capacity to withstand change as to any particular point, would be extremely serviceable.

Then, too, as has probably already occurred to many readers, where Mendel's law applied a breeder might find his best course in working for some special point, to breed first for its contrary or opposite; in other words, to try to make it a "recessive" character, and so determine the individuals possessing it in purity perhaps several seasons earlier than would otherwise be possible.

It is too soon to say yet just where and how Mendel's law applies, but it is certain that it does apply in some cases, and altogether probable that investigation will show it of great use to poultry breeders. To make it useful to themselves poultrymen must familiarize themselves somewhat with it and with investigations along the same lines, and must also do what they can in the way of making observations to discover whether or not it applies with respect to any particular phenomena of breeding.

LESSON III.

Market Duck Culture.

Introductory.

IN this lesson we consider duck culture almost exclusively as it pertains to a single breed of ducks, the White Pekin, which so far surpasses all others in popularity, that market duck culture in this country is White Pekin-duck culture. In our fowls we have in each class a number of varieties, and also have several classes which are either adapted to the same uses, or could without much difficulty be made so; but in ducks we have nothing else that would take the place of the Pekin.

Another peculiarity of modern market duck culture is that it is devoted exclusively to the production of "green" ducks, that is, of ducklings to be marketed at ten to twelve weeks of age. At that age the ducklings have frames almost as large as when full grown, and will dress four to six pounds each, five pounds being about the average weight. Much of this weight is fat, and the proportion of edible meat on a duckling at this age is much smaller than on one of the same weight at four or five months of age, but the profit in duck culture is all in the green ducks, and the duck specialists devote themselves to it exclusively. The older ducks which come to market are mostly from the west and south, grown in small lots on farms, generally under conditions which do not fit them for the green duck trade.

Pekin ducks are much easier to handle in large numbers and in limited quarters than chickens. They grow so much faster that the brooding problem is greatly simplified, and if conditions are at all favorable, and care anywhere near right, they are very free from disease. The common ducks do not grow anything like as fast as the Pekins. Some of the other pure bred varieties may equal the Pekins in growth, and at intervals someone interested in another variety endeavors to start a boom for it, but so far the results have not been flattering. What temporary enthusiasm may be developed does not extend far, and soon dies out. Since the introduction of the Pekin duck no large grower has taken up any other variety, and, I believe, no large success has ever been made with any other duck.

The breeding of Pekin ducks for show and sale for stock purposes receives little attention at present. In the early days of their popularity, when there was a very lively boom in duck culture, poultrymen who went into ducks carried on the duck business on much the same lines as their other poultry business. Some few continue to do so. But the more successful growers of ducks for market generally abandoned the other branches of the business, finding it more satisfactory and more profitable to devote all their time to market ducks. Those who continue to advertise and sell exhibition and breeding stock and eggs for hatching are mostly poultrymen who handle other fowls also.

For those who succeed in it, duck growing is probably the most profitable line of poultry culture, but the field is more limited than the trade in eggs or in broilers, roasters, or fowls. For this there are several reasons. Duck growing on a large scale is a very new industry. It was not until the Pekin duck appeared that tame ducks began in this country to be considered especially desirable for the table. The native duck left to itself frequented streams and puddles that were often filthy, and ate food which imparted strong odors and tastes to both its flesh and its eggs. Any duck left to itself develops the same habits, but the grower of the large improved breeds of ducks finds it worth while to keep them up and see that they are cleanly fed. For the meat of such ducks the demand constantly increases, though the demand for duck meat will always be much less than for chicken, because duck is too rich for a great many, and too expensive for many more. Thus the consumption of ducks is limited, and New York city is probably the only market in the country which can use all the green ducks which might be sent it. Outside of large cities and popular resorts the demand for ducks is light. A poultryman who could easily dispose at good prices of several thousand chickens will find the same market requiring only as many hundreds of ducks.

Hence for most of those who read this lesson the question must be of the production of a few hundred ducks as a part of their undertakings in poultry, and it is on this basis that we will treat the subject, making only occasional or incidental reference to the methods of the large growers. Anyone wishing to start the business in a large way ought to learn it first on a large duck farm. One who begins with a few and raises not more than 200 to 300 at first can get along very well by applying the information here given.

Pekin Duck.

Location for Duck Growing.

For growing the young ducks for market no water except for drinking purposes is needed. The ducks grow faster when kept from the water. For the breeding stock, and for ducks grown for stock purposes, water is not absolutely necessary, but results are generally more satisfactory if the ducks can have access to a stream or to the margin of a pond or lake. Many duck growers who have such a location build the houses near enough to the stream to admit of making yards partly in the water. When the stream is shallow the fences may run right through it. When it is so deep that fences can be used only near the bank the yards may run a short distance into the water. Contrary to the common idea, ducks neither require nor thrive in damp quarters. Though they like to frequent streams and marshy places, they need well drained ground to which they can go when tired of the water, and the house site should be as dry as for hens.

Buildings and Fences.

Houses for ducks are built on the same general plans as hen houses. A building 12 to 16 ft. wide, about 6 ft. high at the sides, and 8 or 9 ft. in the middle, is perhaps the most satisfactory style when many pens are to be kept in one house. For a single pen almost any sort of outbuilding will do, and if a house is to be made especially for the ducks, it need not be other than of the cheapest boards, covered with roofing material or shingles to keep out the wet. It is not necessary that the house should be warm, but it must be dry.

When a number of pens are kept in the same building, it is more convenient to have a passage along the rear wall. The partitions between pens and along the passage need not be

House for Breeding Ducks.

more than 2 to 2½ ft. high, and openings through them are necessary only when the feed troughs are placed along the walk, when a slatted opening in the partition enables the ducks to get at the food in the troughs. When this arrangement is made the trough occupies about half the length of each pen. It is easy to step over these low partitions, and many handle litter and manure to and from the pens over the partitions, but if desired a part of the partition in each pen may be made movable, to admit running a wheelbarrow into the pens. If these partitions are of boards there will be less draft through a long house.

The fences between the yards should be of wire. A fence two feet high will keep the ducks in. Duck growers usually make temporary fences, driving short stakes into the ground and attaching the wire fencing to these with staples, using only two or three at each stake, and not driving them in tight. The fence built in this way is easily taken down and moved, a matter of considerable importance, for ducks foul the ground badly, making it necessary to turn it over and plant on it often, and this can be much better done with the fences up and out of the way.

Number of Ducks in a Flock.

As drakes are not combative as cocks are, flocks large enough to require a number of drakes are kept. The usual plan is to have twenty-five or thirty ducks in a flock. For this number of ducks five or six males are needed, (one to every five females), during the early part of the winter. Toward the first of March the number of males in the pen may be reduced to one to every seven ducks, and in May some breeders still further reduce the males, leaving only about one to every eight or ten ducks. A drake will successfully serve more ducks when the flock has a water run than when only drinking water is provided. All water fowls copulate more freely in water than on the ground.

Ducks kept for breeding should be given good sized yards. They will do fairly well in close quarters, but have not the strength and vitality when so kept that they have if given room to take more exercise. A duck that forages about much is quite strong on its legs, while one that is confined to a small yard and eats only at the trough is very weak on the legs, and will give out after quite a short walk or run. In such condition ducks may lay well, but the eggs will not hatch as well nor as strong ducklings as if the old stock had more strength.

A flock of twenty-five to thirty-five ducks may be kept in a house pen containing about one hundred and fifty square feet of floor space. A little more room will do no harm, especially if the ducks are very large. For the outside yard I would not give specific, or minimum or maximum areas. Give the breeding ducks as much yard room as you can, and if possible let

BEGINNING WITH STOCK OR EGGS.

them have access to water. A single flock on a farm may be given the run of a small field. Where they must be confined make the smaller yards square rather than of parallelogram form, and make them as large as your land will permit. A quarter of an acre of grassy yard makes a nice yard for a flock of breeding ducks of the numbers we are discussing, but if you have the room you may be able to give a considerably larger yard at very little extra expense for fencing.

Beginning With Stock or Eggs.

One of our most successful duck growers and most judicious advisers of new poultrymen says that for those who begin in the fall he thinks it better to buy breeding stock, but those who begin toward spring may find it more satisfactory to start with eggs. From one consideration I would always advise the beginner to buy some breeding stock though not beginning until late in the spring. By handling only a few breeding ducks, and only for a part of the season, he gets some knowledge of them and experience which is of value to him when his young ducks come to their first breeding season. His chances of handling them properly and with satisfactory results are very much better if he has had some experience along that line than if all his knowledge of ducks is what he gained while growing them. It may not be advisable to buy breeding ducks enough at the prices which must be paid in the spring to hatch a large lot of ducklings, but I certainly think it will pay anyone not familiar with the nature and habits of ducks to buy at least a trio of them, though also buying eggs to hatch.

What Kind of Stock.

In buying stock ducks buy good ones. That is, ducks that are good for market purposes. They should be of good size, for to be profitable ducklings must make pretty nearly five pounds on the average at ten weeks, and such ducklings cannot be produced from small ducks. W. R. Curtiss & Co., who for years have bred Pekin ducks for all purposes very successfully gave the following statement of their methods of mating in FARM-POULTRY a few years ago:—

" We select females of good fair size,—we like to have them weigh at maturity eight pounds

Brooder House and Runs for Ducklings.

each—medium long in body, deep at keel, with flat backs and short necks; we also see that they have a sharp, quick eye, and carry themselves well when walking.

"If we wish to produce large birds we use rather small, active drakes, weighing, say six to seven pounds each. We do not use small drakes constantly year after year, for that would run the stock out. Our favorite mating is of medium sized birds of both sexes—the drakes to be as active as possible. This produces good fair sized stock, what we are after for market.

"Another mating we use quite frequently is, drakes weighing eight or nine pounds with extra large ducks weighing ten to eleven pounds each. This mating has produced for us extra large ducks of both sexes."

Feeding the Breeding Stock.

Ducks are fed practically all soft food. A number of growers have at different times tried giving hard grain as to hens, but generally the result has been to curtail the production of eggs. Herewith are given the rations for breeding stock recommended by some of the leading growers:

Ration I. "Equal parts of corn meal, wheat bran, and low grade flour, with about twelve or fifteen per cent of animal food. One-fourth of this food should be composed of vegetables cooked — say, small potatoes, turnips, etc., with all the green rye and refuse cabbage they will eat. Feed this mixture mornings and evenings, giving a little corn, wheat, and oats at noon. Feed all the birds will eat up clean and no more."— JAMES RANKIN.

Ration II. "Equal parts corn meal and shorts with ten per cent beef scraps added. If green food is not available, add one-fifth cooked vegetables to the mash. Give raw vegetables at noon two or three times a week."—WEBER BROS.

Ration III. "One-fourth corn meal, one-fourth bran, one-eighth broken crackers, one-eighth gluten meal, one-eighth low grade flour, one-eighth beef scraps. This mixture makes two-thirds of the mash, the other third being scalded green clover cut fine, and boiled potatoes mashed. The grain feed is mixed dry at first, and then the clover and potatoes added, and the whole mixed with hot water and fed warm. Feed all they will eat up clean with a relish. Allow no food to stand before them at feeding times." — CURTISS BROS.

Ration IV. "Two parts bran, one part Indian meal, two parts cut clover or other green food. Ten to twelve per cent of the whole to be best quality of beef scraps." — POLLARD.

Ration V. "Two parts clover heads, boiled, two parts corn meal, two parts middlings, two parts bran, one part ground bone."— MCFETRIDGE.

The food for ducks may be fed either cooked, scalded, or simply wet with warm or cold water, but it is advisable to use the same method continuously, and not indifferently and indiscriminately.

It is of greatest importance that both animal food and green food be liberally supplied. Ducks are hearty feeders, and, as when laying a duck lays almost daily, she must be well fed and with a good substantial ration. Water for drinking must be constantly accessible, and it should be near the feed troughs, for ducks wash their food down with water, and may choke if not provided with it. Grit and shell should be constantly accessible. Ducks can be started laying quite readily in early winter by feeding meat heavily in the mash, and by keeping them confined to the house on raw and stormy days. They will stay outdoors by preference, even in cold stormy weather, but may not begin laying if allowed to follow their inclination in this.

The floors of the pens should be bedded with coarse hay, straw, or leaves. Nest boxes are seldom used. The duck will scoop out a nest for herself in a corner or at the side of the pen, and duck growers agree that eggs are less likely to be broken when the ducks are left to themselves in nesting than if nest boxes are provided. Ducks lay very early in the morning, mostly before daylight, though occasionally a duck will lay as late as eight or nine o'clock in the morning. If they have liberty they are just as likely to drop the eggs about the yards as to leave them in nests in the houses. If they have access to water they will oftener drop eggs in the water where they are lost. For this reason ducks that have access to water should be kept from it in the morning until the eggs for the day have been laid.

Hatching the Ducklings.

Where only a few scores to several hundred ducks are hatched hens or incubators may be used as most convenient. For larger numbers it will usually be found more satisfactory to use the incubators exclusively.

When the eggs are hatched under hens give nine to eleven eggs to a medium sized hen. The number a hen can cover properly depends on the size of the eggs, and also on the season. If eggs are very large it may be better not to give an ordinary sized hen as many as nine early in the season. After the weather is warm a hen may be given a nest full, and often hatches just as well when there are more eggs than she can fully cover, but in the winter and early spring it is better to limit the number to what it is clear a hen can keep warm all the time, for when there are too many eggs in the nest all are likely to be chilled in turn and the entire lot spoiled, when with a few less eggs in the nest a good hatch of strong ducklings would have been secured.

When duck eggs are hatched in incubators the operation of the machine is varied in some makes. In other machines a special pattern is made for duck eggs. The poultry keeper who has a machine adapted to both kinds of eggs can learn from his book of instructions or from the manufacturer or agent what special adjustments of it to make for duck eggs. In buying one can ascertain to what extent a machine under consideration is adapted to duck hatching.

The period of incubation for duck eggs is twenty-eight days. Sometimes the ducks hatch earlier, but it is preferable to have them go the full time. Unlike chickens, the little ducks pick the shells quite a long time before they attempt to break out, usually thirty-six to forty-eight hours. One inexperienced in duck hatching sometimes becomes alarmed about the ducks not coming out promptly after breaking the shells and undertakes to assist them. There is no occasion for this. As the ducks pick the shell so long before leaving it it is necessary to see that the membrane does not have a chance to dry and adhere to the duck or become so tough as to interfere with the hatching. In running an incubator the ventilation is reduced so much at the last that the air in the egg chamber is saturated with water and the membranes kept moist. In hatching with hens make sure that the hens are perfectly comfortable, and nests and hens free from lice. Then the hens will not be likely to leave the nest for long at a time. It is well also to keep the hens confined to the nest as closely as possible after the eggs are pipped, though as so long a time may elapse between the pipping of the first egg and the exclusion of the last duck the hen should have at least one opportunity to leave the nest in that time. The best way is to let or take the hen off just about the time the first duck is due to push out of the shell; then keep her on the nest until the hatch is over. It should be observed that the period of incubation is a full week longer than for chickens, and that this week, if the sitting hen is not in good condition, is harder on her than all the rest.

Brooding Ducklings.

When only a very few ducks are hatched at one time, say, not more than two or three hens can brood, hens may be satisfactory mothers, but as the number in a lot approaches the capacity of a single brooder, it will be found much the more satisfactory way to use a brooder, or as many brooders as are required. I think duck growers generally prefer to use the pipe system of brooding if they have enough ducklings to use such a system. The brooder house for ducks may be just the same as for chickens.

The ducklings are not taken from the nest or machine until they are twenty-four to thirty-six hours old. If they are to be brooded by hens the hens should be confined so that they cannot roam about and wear out the little ducklings. After the weather is warm the ducklings may be allowed considerable liberty, provided they are not disposed to wander too far, in which case they should be restrained. Sometimes little ducklings but a week old, if free to do so, will stray long distances. This must not be allowed if good growth is desired. It is a very simple matter to confine the ducklings to any required spot, for a board ten or twelve inches wide, set on edge and supported by pegs driven beside it will keep them in until they are some weeks

LESSONS IN POULTRY KEEPING — SECOND SERIES.

old. This is easily moved as often as desired. For a more permanent fence it is better to use wire netting. When the ducks are very small the finest mesh may be required to keep them in, but they grow within a few days to such a size that the two inch mesh will answer. Which to use must depend on the amount used, and on the convenience or inconvenience of changing the quarters of the ducklings, or changing fences.

Temperature of Brooder.

The brooder into which the ducklings first go should be at a temperature of about 90° before the ducklings are put into it. The heat of the ducklings will raise the temperature several degrees. This temperature, approaching 95°, is about what the ducklings should have for the

Pekin Ducklings Six Weeks Old. Boy Six Years Old.

first day or two, it being a reduction of about 8° from the temperature of the incubator. The temperature should be gradually reduced until at the end of a week it is 80° to 85°. Ducklings grown in the winter need artificial heat until ready for market. For those hatched in the latter part of winter, and in the spring, the period of brooding is shorter. When settled warm weather comes the ducklings need artificial heat only for two or three weeks. Ducklings that are five or six weeks old about April 1st can go into "cold" brooder houses at that time. These houses are tightly built, so that the ducks in them are well protected, but have either no heat at all or merely a row of pipes along the rear wall, perhaps a couple of feet from the floor to take the chill off the house should the weather at any time be unusually cold.

Feeding the Ducklings.

The very first thing the novice in duck culture should fix in his mind about feeding ducklings is that the duckling must have "grit" in its food from the very start. I am not prepared to say whether it would be possible to have thrifty ducklings without grit, but I have tried a good many ways of preparing their food, and found that they always needed some grit in the beginning, and that they did best when it was mixed with the food. Any fine grit or clean coarse sand will do, and not a great deal is required. When mixing a pailful of mash

FEEDING RATIONS FOR DUCKLINGS.

for ducklings, I would throw in a heaping handful of fine grit or coarse sand. Mixing lesser amounts would use grit in proportion. Some duck growers use grit in food regularly for both young and old ducks; others only put it in the food for the young ducks for two or three days. I have never discovered any advantage in continuing to mix sand or grit with the mash after the first few days, unless symptoms of indigestion appeared, when the use of these aids to digestion for a few times seemed decidedly beneficial. Grit should be accessible to the ducklings at all times that they may take it if they need it.

As for chickens, there are many good food combinations for ducks; perhaps not so great a variety, for they are fed practically all soft food. Herewith are given the rations used and recommended by a number:

Ration VI. "Two parts wheat bran, one part meal, with a handful of fine gravel or sand thrown in, mixed with either hot or cold water, or with skim milk, to a crumbly consistency. I frequently break raw eggs into the grain in the proportion of two eggs to one quart of the dry grain. This must be thoroughly mixed that it may not be pasty or sticky. After the first three days omit the sand or gravel, and by the fifth day begin to feed a slight proportion of beef scrap, increasing gradually until at two weeks of age they are getting five per cent beef scrap. Gradually increase the animal matter until at five weeks of age the ducks are getting fifteen per cent of it, which proportion may be continued until killing time."— POLLARD.

Ration VII. "Equal parts corn meal, bran, flour, ground graham bread, and rolled oats, five per cent beef scrap, a little oyster shells and grit, and a little finely cut green rye. Moisten with cold water. The ingredients are first mixed dry, then water thrown on and mixing continued until the food is of the proper consistency. Feed this five times daily until the ducklings are three weeks old.

"After three weeks feed two parts corn meal, one part bran, one part middlings, ten per cent of this amount beef scrap, a little oyster shell, and finely cut green stuff. This is fed until the ducks are six to seven weeks old, when they are put on a 'fattening ration' composed largely of corn meal, just flour enough being added to hold it together when wet; the proportion of beef scrap being still about ten per cent."— WEBER BROS.

Ration VIII. "*First week* — equal parts of corn meal, middlings, crackers, or stale bread, and green stuff; mix in a small handful of sand to each quart of food. Give occasionally bread soaked with milk for a change.

"*Second week* — four parts corn meal, two parts wheat bran, two parts middlings, one part beef scrap, sand; mix with about one-third the quantity of green stuff. At about six weeks put ducks in fattening pens and feed two-thirds meal, the remainder about equal parts of bran, middlings, and greens; add about twelve per cent of the whole beef scraps."— HALLOCK.

Ration IX. "*First four days* — four parts wheat bran, one part corn meal, one part low grade flour, five per cent fine grit. Feed four times a day what they will eat up clean.

"*From four days to three weeks* — three parts wheat bran, one part corn meal, one par — low grade flour, three per cent fine grit, five per cent fine beef scrap, soaked. Finely cut green clover, rye, or cabbage. Feed four times a day.

"*From three to six weeks old* — equal parts corn meal, wheat bran, and low grade flour, ten per cent beef scrap, three per cent grit. Feed three times a day.

"*Eight to ten weeks old* — one-half corn meal, equal parts bran and low grade flour, ten per cent beef scrap, three per cent grit; oyster shells and less green food. Feed three times a day.

"The above ingredients should be made into a mash, and should be crumbly, not pasty. Proportions by measure, not by weight."— RANKIN.

Whichever of the above methods of feeding the reader takes up, he will find it better to follow that method as closely as he can, including the accessories perhaps not insisted upon in other rations. The different proportions of different articles used by different experts, practically balance each other, as one who is skilled in feeding can see.

If only a few ducks are kept on a plant on which fowls also are grown it is not necessary to make a special mixture for the ducks. The mash made for the hens will answer by the addition of a little sand or grit, when required, to the portion needed for the ducks. When the ducklings are fed this way, extra green food should be provided.

Importance of Water.

Even more than the old ducklings, the little ducks require water to wash down their food. Their drinking vessel should be always supplied. They will drink even at night. For a small brood of ducks with a hen an open drinking pan or saucer may be used, but for larger lots drinking fountains into which the ducklings cannot get should be used, for with their pushing and crowding at feeding times a brooderful of ducklings having access to an open drinking vessel get themselves and each other very wet, and sometimes the wet ducklings are chilled or trodden down by the stronger ones. Milk may be used for mixing the food, but should not be given to drink, because the ducks will become smeared with the milk, which drying on them puts them in a most miserable condition.

Cleanliness.

It is important that all feeding and drinking vessels should be kept clean. The duckling's voracious habit of eating makes it shovel down filth with its food if there is filth present, and in a very short time the effects of sour and foul food and drink become apparent. The floors of the pens should be littered with hay or planer shavings, and as these become wet and soiled they should be removed and fresh litter put in. The yards, when small, should be swept or scraped, say once a week, and even if a yard is large enough to make cleaning of the entire space unnecessary as often as this, the parts of it most frequented by the ducks will need frequent cleanings.

Ducklings Must Be Kept Quiet.

Ducks are timid fowls, and the Pekin duck is probably the most timid of domestic ducks. Any annoyance or disturbance of the growing ducklings (or of the breeding stock) is therefore to be avoided. Visitors should be kept from the duck quarters as much as possible, and dogs or other animals that would frighten the ducks must be kept off the premises. Sometimes ducks become panic stricken at night and rush back and forth in the house or yard, wearing themselves out, losing a great deal of weight, and checking growth. To keep them quiet breeders who have trouble with them in this way hang large lanterns in the duck houses.

Marketing Ducks.

The well grown and well fed duckling should be ready for market at ten weeks of age. Early in the season some growers market ducklings at nine weeks to get the extreme high prices for the first ducks; but as soon as the ducks begin to go to market in fair supply it is better to hold them until at their best. For the bulk of the lot this should be at ten weeks. Some of the inferior ones will not be ready for a week or two more. All should be sold by twelve weeks, for after that the duckling begins to grow its adult plumage, loses its "baby fat," and loses weight, and will not be fit to kill until eighteen to twenty weeks old. At that age the duck is meatier and better eating, and is far more satisfactory for the home table, but will not bring as much in the market as the green duck, and will have cost nearly twice as much.

The large duck growers, from the time their first ducks are ready to dress until the close of the season, have pickers constantly at work. Their product goes almost wholly to wholesale dealers and commission houses. The grower who has only a few hundred ducks, and has a trade in dressed poultry, should have no trouble in disposing of that number of ducks at good prices to his trade.

Killing and Dressing.

The methods of killing, dressing, cooling, and packing ducks for market are the same as given for poultry in Lesson XX., First Series, but a duck is much harder to pick by any method than a chicken. It is a slower job and one that requires patience.

Growing Ducks for Stock.

The ducks that are to be reserved for stock purposes should be reserved from either the early spring hatches or from lots hatched not later than May. Good ducks often come from later hatches, but systematic selection from the best lots is the only way to keep up size and stamina. The best duckmen select what they need of the finest April and May ducklings, and separating them from the market lot at the age when fattening begins, continue them on such a ration as has been used for growing them, at the same time giving them more room with access to grass and water range. As they approach maturity the proportion of meat in the ration is reduced. Often it is completely eliminated for a time; because if continued it gets the ducks to laying earlier than is desirable. A "maintenance ration" only is given through the fall until the ducks are put in breeding houses at the beginning of winter, when rations for breeding stock, as given earlier in this lesson, are given them. It is not desirable that the ducks should lay much before December

LESSON IV.

Goose Culture.

THIS is the branch of poultry culture which interests the least number of people. You may travel long distances in many directions, and see very few geese. Yet it is a line that might profitably be given some attention by practically all farmers and a great many others.

I do not mean by this that the growing of geese on a commercial scale could be carried on profitably very much more generally than it is. No doubt interest in goose culture on a commercial scale could develop considerably before an over supply was created, but those to whom

Embden Goose.

goose growing should appeal most at present are the many who might grow a few geese every year, either for home use or for a local market, almost without expense. Almost every farm has some land not used for any other purpose, and not productive, that might be devoted to goose growing. On almost every farm a flock of geese, large or small, according to accommodations, might be kept on pasture with other stock.

Geese thrive most readily and are least trouble to handle, and grown at least cost, when given good pasture, but they may be grown like ducks in close quarters if green food is liberally provided. This way of growing them is not recommended. I merely mention it as possible for those who would like to know something of geese, but have not good natural facilities for growing them.

Under favorable conditions geese are the easiest of all domestic fowls to handle. With good pasture provided with water they may be left to themselves half or more of each year. They do better, generally, if provided a little grain food to supplement what they get by foraging, but it is not regarded necessary, and perhaps the majority of those growing geese give the old stock nothing during the pasture season, and feed the goslings only a few weeks.

The Breeds of Geese.

Many of the geese found in this country are the common gray, gray and white, or white geese, not much larger than a large duck, and in a general way showing the same inferiority to the improved breeds of geese that common fowls of all kinds do to the well bred.

Of pure bred domestic geese there are the following breeds: Toulouse, Embden, African, and China, or Chinese. Of the last named breed there are two varieties: Brown and White. Canadian, or Wild Geese, are also kept in captivity, and bred to be sold as decoys to hunters. In some sections the breeding of Wild geese for this purpose is carried on in a small way by many persons, and considering investment and attention required is quite profitable. The young geese generally bring at the hunting season in the fall $5 each.

Besides being used for the purpose just mentioned, Wild geese in captivity are crossed with domestic geese, producing a goose which is properly a hybrid, and sterile, but which is commonly called a Mongrel goose, and perhaps oftenest called the "Rhode Island Mongrel goose," most of them coming from that state. If the reader will make a mental note of the way in which the word "mongrel" is used in connection with geese, he may save some confusion on the subject. The mongrel, or common fowl of the barnyard, is an indiscriminate mixture of blood like the common goose, but the Mongrel goose is a half blood wild goose.

Of the breeds of domestic geese the Toulouse is a very large gray goose; the Embden a very large white goose; the African a very large brown goose, the "Standard" weights of all being the same for adult specimens, viz., 20 lbs. for males and 18 lbs. for females. For young specimens the weights are: Toulouse — gander, 18 lbs.; goose, 15 lbs. Embden — gander, 18 lbs.; goose, 16 lbs. African — gander, 16 lbs.; goose, 14 lbs. Why these variations in the weights of young geese should be made I do not know, nor do I think they can be given a reasonable explanation; but it makes little difference to any goose growers but the few who exhibit. Chinese geese are much smaller than the others, the weights being: Adult gander, 14 lbs.; adult goose, 12 lbs.; young gander, 10 lbs.; young goose, 8 lbs.

Of these varieties the Toulouse and Embden are quite common, flocks of them being found in almost all parts of the country. The others are more rarely seen, though Africans are quite popular in Rhode Island, and occasional flocks or specimens of the Chinese may be met in any part of the country.

Though similar in size these breeds of geese differ in qualities and characteristics. I can, I think, present these differences in no better way than by quoting the opinions of two of the best authorities on the subject:

Mr. Samuel Cushman, in an article written for FARM-POULTRY some years ago, said:— "Toulouse geese are very large, docile, and great layers for so large a breed; but they are slow to mate; ganders mate with fewest geese of all, and their very early eggs are rarely fertile. They grow a very large frame, but grow slow, and do not get plump or fatten readily until late in the fall. Their plumage is easily plucked, but as it is dark their skin does not have the attractive appearance of that of the white goslings. Therefore, this breed is not just suited to the production of early goslings.

"Embden geese are large, docile, and their goslings grow plump and large very quickly; have yellow bills and legs, and the most desirable white plumage. They also pick easily. They seem almost all that could be desired of any pure breed, but they have their faults. They lay later in the season than any other breed, and fewer eggs. The ganders are also backward in mating, and many infertile eggs are produced. The goslings that are secured are all that could be desired, but they are few in number and appear on the scene rather late.

"African geese are as large as the Embden and Toulouse; lay more eggs than the Embden, and nearly as many as the Toulouse, and lay them nearly as early as the Chinese varieties. They are prompt to mate, and the ganders will mate with more geese, and their early eggs will be more fertile than those of any other large breed. They are steady, docile, and great feeders.

Toulouse Geese.

Goslings grow faster up to the time they should be marketed, and are easily fattened. Many more large goslings will be secured early than from the other varieties. If you have ten females each of Africans, Embdens, and Toulouse, with the proper number of ganders on your place, and give them the same chance, you will have after the hatching season is over many more African goslings than either Toulouse or Embden. Probably three times as many Africans will be secured as Embdens, and twice as many as Toulouse. This has been our experience. Of course there are exceptions to all rules. African goslings are more salable dressed than anything except the Embden, and as they get in ahead, bring as much or more per pound. Their drawback is their dark bill and dark plumage."

Of the Chinese geese Mr. Cushman expresses this opinion:—"They are small, nervous, and noisy by nature, and are the least able to withstand cold or heat. They are timid and excitable, and hard to fatten, also to pick, but they are the earliest and most profitable layers and producers of all. Their pure bred goslings are the least favored by fatteners and marketmen of all varieties."

Mr. Geo. H. Pollard, comparing the breeds of geese, says:—" There has been considerable question as to which is the best breed for general purposes. The three breeds which have the strongest following are the Toulouse, African, and Embden. Each of these has its special virtues. The Toulouse is the largest of the three, and produces a very strong and quick growing gosling. Owing to its great size, however, it is longer in reaching maturity, and when fully matured is too large to suit the best trade in the New England market. The female produces a fair number of eggs, and, on the whole, the breed is easily managed and is profitable.

" The African is a smaller bird, is darker in plumage, and enjoys the distinction of being the hardest to dress of any of the breeds. The female is the best layer of the three breeds mentioned. The eggs are of a fair size, and under proper treatment are generally well fertilized. The young are strong and hardy, and grow quickly to a size which is well suited to the market demand. The principal objection to the young of this variety comes from the fact that they are at most stages liberally covered with colored pinfeathers and down, which serve to mar the value of the carcass for high class family trade.

" The Embden unites the good qualities of the Toulouse and the African, while it is without their failings. The mature birds are a beautiful clear white, with flesh colored bills and legs. They are of medium size and weight. The females are good layers. The young are strong and quick growers, and being without dark pinfeathers and down, they dress clean and white at any age. The carcass is more tempting and sells more quickly to fastidious consumers. The quality of the flesh is about the same in all three. Probably no one could detect any difference in either tenderness or flavor. The Embden or Toulouse are more tractable and easy to manage than the African, which is the most pugnacious of the three breeds."

From these opinions, which, though not strictly agreeing are not radically different, the reader may judge something of what breed will be most satisfactory to him. Messrs. Cushman and Pollard are, of course, considering the three breeds as producers of goslings to be sold at periods of high prices or to an exacting trade. The reader who wishes to produce for himself, or for a local market, and does not intend to go into geese extensively, may find faults that were serious from their point of view of little importance from his. Mr. Cushman advocated very strongly the use of crossbred goslings for market. It should be said, however, that the experiments made with crosses took account of too small numbers, to give force to the general conclusions drawn from them, and that very few goose growers have done anything systematically in this line except those engaged in the production of the "mongrel" geese. In the goose growing sections of Rhode Island where geese are bred for market more extensively than anywhere else in this country, the geese used are generally grades produced by various mixtures of the blood of the three improved breeds we have under consideration with the old common goose. The goslings from this stock, while not equal to the best of the pure breeds or their direct crosses, grow rapidly to a good size and seem to suit the requirements of the trade.

Getting a Start With Geese.

If one wishes to grow more than a few goslings year after year the best plan is to begin with adult birds three years old or over. Geese do not come to full maturity until about three years of age, and rarely give at all satisfactory results until two years old. The young geese will lay and some of the eggs may hatch, but the goslings are apt to be weakly and not thrifty.

Good stock of any of the improved breeds usually costs $3 to $5 per bird. It is better to purchase some months in advance of the breeding season, as the geese will not breed well if moved just prior to the laying period. If stock has not been purchased early it is usually better either to let the matter go over for another season, or to buy eggs. In any case one must expect it to take several seasons to get a flock of geese established in new quarters, and breeding satisfactorily, for good breeders of mature age are not often offered for sale, and the young geese require two years, and sometimes more, before they produce well. Once established, however, a flock of breeding geese can be kept unchanged for quite a long period. The females are said to be profitable up to ten or twelve years of age, and males to a little more than half those ages, say, six or seven years.

From one to four females are mated with each male. Young geese are disposed to pair. As the ganders grow older they will usually serve more mates, but they do not copulate indiscriminately as fowls and ducks do. Each gander serves only his own mates, and an unmated goose in a flock in which all the ganders have other mates may be entirely neglected. For this reason goose growers often put an extra gander with the flock to pick up, and mate with any neglected or discarded females. Usually when a gander and goose have mated once they are faithful to each other as long as allowed to remain together, and often refuse to take another mate for a long time after being separated. This trait makes it possible once the geese are mated to allow as many families as desirable to run in one flock, even though they be of different breeds.

Houses and Fences.

Geese require very little shelter, and the heavier breeds are restrained by almost any sort of low barrier. They do not seem to mind severe storms in the least, but will lie down on the snow, draw their feet up into their plumage, and seem supremely contented, no matter how

Flock of Breeding Geese in Pond.

cold or rough the weather is. I have some this winter that have gone into a small shed with a cockerel for which they seemed to have a special liking, but as soon as the cockerel was taken away quit going under cover and remained out in severe sleet storms. A low rough shed open to the south will provide all the shelter needed, and it need not be much larger than will give the flock standing room under it. My own experience with geese has been limited to small numbers, but though I have always provided shelter for them, and kept the floors littered the geese have made so little use of these comforts that continuing them seems more in the line of satisfying my conscience than providing for their needs. Sometimes they have to be confined at night for protection, but, as a rule, they can take care of themselves.

Care of Breeding Geese.

Geese breed best on pasture with streams or ponds accessible. They require little grain if the pasture is abundant. If pasture is short it may be supplemented by grain and vegetables. In winter they should be fed mostly on vegetable food, cut clover, alfalfa, cabbage, beets,

CARE OF BREEDING GEESE — HATCHING GOSLINGS.

turnips, etc., with oats or barley and bran. Corn is generally too fattening, and most breeders are very sparing in feeding it. Herewith are given the rations of several well known breeders of geese:—

Ration I. "Through the laying and breeding season, in addition to grass, they should be fed twice a day with shorts and corn meal, equal parts, thoroughly moistened with cold water, but not too wet lest it produce diarrhea; add ten per cent of beef scraps, or its equivalent."— RUDD.

Ration II. "Take some boxes about eight inches deep, and put oats or barley in them. Place these in the pasture away from other fowls. There is no danger of overfeeding grain if the geese have pasture. Give corn only in coldest weather — when it is storming, or there is so much snow that they cannot go foraging."— NEWMAN.

Ration III. "Turn out on pasture from June until fall; feed no grain while grass is available, then feed lightly of oats and whole corn. After February 1st, give full ration: A mixture of corn meal, shorts, beef scraps, boiled potatoes, or turnips in the morning; whole grain in the afternoon."— WILBUR.

Geese usually begin laying late in the winter, or in the early spring. Not many lay while there is snow on the ground. When the goose is ready to lay she is apt to seek out a nest for herself, and having once selected a place is impatient of any interference with her laying habit. The goose, as a rule, looks for a somewhat secluded place, and as a pasture often furnishes few such places, keepers make a practice of providing nests that may attract the geese. An empty barrel placed on its side with a little earth in it to hold it steady and make a foundation for the nest, then a little straw or hay added, is the usual nesting place furnished. Often the geese will ignore these and make their own nest right in the open.

As geese lay but few eggs it is customary to take them away as laid during the first laying period, and give them to hens to hatch, or place them in an incubator, and not allow the geese to incubate until they have laid a second lot of eggs.

Hatching Goslings.

The hatching of goslings by natural methods, whether with hens or geese, differs little from the hatching of other fowls. The points of difference in hatching with hens are those which depend upon the size of the eggs and the length of the period of incubation. The eggs of the large breeds of geese are very large, and early in the season three will be found enough for a medium sized hen. Later when the atmosphere and ground are warmer more eggs may be given, but it is rarely wise to give more than five large goose eggs to a hen. The period of incubation being thirty days, approximately half as long again as the incubation period for hens' eggs, hens that are not in good condition and free from lice are likely to neglect or desert their eggs. Hence besides taking special care to use hens that are in good condition and to guard against lice, which multiply much faster on a hen late in a prolonged period of incubation than while she is active and robust, it is well to watch the hens very closely after the third week of incubation, and remove any that seem to be losing condition, giving their eggs to fresher hens.

When geese are used for hatching they may be allowed twelve to fifteen eggs. Usually they must be set where they have been laying. Many growers simply leave the eggs of the second litter in the nest.

Goose eggs can be tested from the fourth to the sixth day, and the infertiles removed. Testing should always be attended to with the early hatches and under hens. With eggs set under geese late in the season it is not so essential, for the eggs usually run very fertile then, and as the incubating goose does not like interference it is as well to let her alone.

Artificial Hatching of Goslings.—Artificial methods are very little used for hatching goose eggs. Only a few operators of incubators have had at all satisfactory results with them. One of these, Mr. Samuel Cushman, gave me a few years ago the following points on the artificial incubation of goose eggs:—

"Goose eggs need more drying out than hen or duck eggs, and also need more cooling. When an incubator containing goose eggs is run as for hens' eggs the air cells are too small, not enough of the fluids of the eggs are absorbed, the goslings are glued up, weakly, and cannot get out of

Goslings at Pasture.

their shells. Goslings under hens are also much slower getting out than those under geese. Sometimes they remain for a whole day after pipping the shell, with bill protruding through the opening, gaining strength before making a final effort. When in such a state, if the air in the egg chamber of the machine, or the air surrounding the nest of the hen is not moist, the membranes of the egg will adhere to the gosling so that it cannot get out. After the first week, whether under hens or in incubators, they need moisture. If in nests they should be sprinkled every two or three days or put in water from thirty seconds to a minute. With goose eggs in incubators, especially in ordinary hot air incubators, it is very important to soak the eggs during the latter part of the hatch. They may be soaked every two or three days after the fifteenth day, or daily during the last ten days.

"The temperature required for the incubation of goose eggs is lower than for hens' eggs, but the machine need not on that account be run at a lower temperature. Required conditions are met by running machine at usual temperature, and cooling eggs more, thus giving them a lower mean temperature. Toward the last the heat will naturally run up, and it is best to allow it to do so. Some of the best hatches were obtained when the thermometer registered 106 and 107°. To furnish the required moisture while the eggs are hatching pans of hot water are put for short periods beneath the trays. Under such conditions the largest per cent hatches were obtained. This process, however, must be conducted with caution, for it is easy to overdo it, and then the dried matter within the shell seems to dissolve and glue over the gosling. What is wanted at the period of exclusion is a high degree of heat not too long continued, and just moisture enough to prevent the membrane drying and sticking while the goslings are working their way out.

"The cooling in the incubators should begin at the tenth day. It need not at first be very carefully done, a slight cooling sufficing; but as the animal heat in the fertile eggs increases more care must be taken, and after the eighteenth day cooling and moistening should be very carefully attended to until the eggs pip. Moistening the eggs cools them rapidly by evaporation of the water on the shells, and it is possible that the dipping in water is more beneficial in cooling the eggs than in furnishing the required moisture, for the cooling effect is always pronounced, while the moisture effect of a single application of water is imperceptible. When cooled dry, the cooling toward the close of the hatch should continue until temperature has been reduced to 80 to 85°."

FOOD AND CARE OF GOSLINGS.

After the Goslings are Hatched.

Whatever method of hatching is used, the goslings should be kept warm and quiet after exclusion, and for the first few days. Mr. Cushman thought it better to remove goslings at once from the egg chamber and place in brooder. With hens it is well to remove the goslings as hatched to a flannel lined basket kept in a warm place, until the hatch is completed, when they may be given to the hens that are to brood them. It is best to remove the hen from the nest, and give her a box large enough to let her turn about and give the goslings a chance to get from under her and move about a bit, yet not get away and get chilled. With a goose mother the goslings may be let alone, and as they are generally much stronger when hatched by a goose, are not so likely to suffer from slight exposures.

Except for the precautions to be taken during the first few days, the care of growing goslings is as simple as anything can be. I give here first statements as to care and feeding, from several of our best authorities, and following these a few comments from my own more limited experience, and from observations made in sections where the growing of geese is generally carried on.

Food and Care of Goslings.

Ration I. "The first two or three days keep them in a warm place and give a little soaked bread and water. In nice weather turn them out in small enclosures which can be moved every day. After a week let them go. The first four or five weeks give nothing but stale bread occasionally, but always leave them at liberty to get all the grass or clover they want. Do not soak the bread, as they do not like it so well. After five weeks give a mash of two-thirds bran and one-third corn meal. To fatten — after six weeks, feed one-half bran, one-half corn meal; do not feed it sloppy. Never allow goslings to go to the water until fully feathered, and then only let those go which are to be kept for breeders."—NEWMAN.

Ration II. "Goslings are better off if they get nothing but tender grass and water the first day they are out, or before they are forty-eight hours old. The next day they may be fed two or three times, but very lightly, with scalded cracked corn. This is probably as good food as can be given from then on, provided they have at all times an abundance of tender grass to eat, and the amount of cracked corn fed is such as will leave them always hungry for grass. An exclusive diet of grain or dough without plenty of grass, or too great a quantity, even with grass, will spoil them—cause them to lose the use of their legs, and die. If grain is fed sparingly while they are young, grass being three-fourths of their food, few will be lost. To make the best growth they should have succulent green food before them while they can see to eat. If shut in for a short time morning or evening, or on a stormy day, they should have a continual supply of freshly mown rye, oats, clover, or corn fodder. Otherwise they will fret and lose much weight. When the object is to raise show birds of great size and frame at maturity, it may be best to feed oat meal, gluten feed, and bran liberally, as well as corn and grass or clover; but there will be less uniformity under this feeding, and more will be lost."—CUSHMAN.

Ration III. "They need plenty of green food; and soft tender grass, rye, or other growing grains should be freely fed. A good green run is really one of the necessary aids to the proper growth of young goslings. While many of the old timers disbelieve in the feeding of grain to young birds, we strongly urge that they be fed regularly three or four times a day, as much as they will eat of a mixture similar to that which is fed to the young ducks. This, in addition to the green range will cause a growth that will astonish the natives, and yield a profit to the producer. The young, as well as the old, need free access to stores of grit and shell. While not so timid in disposition as ducks, they should be kept free from disturbance, and all possible pains taken to keep them contented and happy; contentment means growth, and growth means profit."—POLLARD.

My own experience in growing goslings favors Mr. Pollard's methods rather than the others. Feeding whole or cracked grain freely, even as part of the ration, I have found likely to put goslings off their legs, and I have invariably brought them right up again by going promptly back to a mash feed as soon as signs of this trouble appeared. If grain is sparsely scattered the goslings take a good deal without injury, and after they are well grown will eat a good deal of hard grain and grow very fat on it without showing any indications of weakness.

In visiting the Rhode Island growers of geese, one may find great diversity in methods of feeding as far as the use of foods supplementing the abundant pasturage given is concerned. I have seen the goslings on one farm fed a stiff dough largely of corn meal, those on the next a mixture of much the same consistency, but carrying less meal and more bran; while perhaps at the very next place visited the food would be more of the consistency usually given hogs to drink; yet under all these different methods of feeding the results appeared to be equally good, and from this I think it reasonable to conclude that with good pasture the kind and consistency of grain food makes little difference.

One point especially necessary to observe if the goslings are to be kept healthy is to either give pasture large enough to have clean grass in abundance, or move the geese often. A favorite method with those who grow them in large numbers is to divide a small field into three or four sections, and pasture one section at a time.

Marketing Geese.

For profit geese are sold at ten to twelve weeks old, at which ages they should weigh from 8 or 9 to 12 or 13 lbs. If pushed right along they should be ready to market at ten weeks, and at their best weight as green geese at that age. If intended for home consumption they may be kept longer, and as most people want them to eat in winter it may be as well in growing for home use to feed sparingly of grain foods until near the time of killing, when grain may be given freely to fatten.

Methods of killing, dressing, etc., for market are practically the same as for ducks.

LESSON V.

Turkey Growing.

THE turkey is the one common domestic fowl that seems to be utterly unadapted to confinement. The guinea fowl has much the same roving disposition and fretfulness under restrictions, but guinea fowls are by no means common. Geese, as we have seen, are much easier grown when given freedom and good pasturage, yet they may be grown without much difficulty in confinement. Instances of turkeys successfully grown in close quarters are extremely rare, and most of the successful growers of turkeys give them practically unlimited range.

The bulk of the turkey crop of this country comes from the large grain growing and general farms of the central west and southwest. There is a very limited area in Rhode Island where most of the farmers grow turkeys, and some grow a good many. In Vermont a good many turkeys are grown. The other New England states produce practically none, while through the Middle states the turkey product is light and scattering. In the grain growing sections of the west and southwest, turkeys, though not as generally kept as chickens, are very abundant, and probably more than nine-tenths of the entire turkey product marketed comes from these sections.

On these farms a flock of turkeys will forage far and wide, and if not too large, will pick the greater part of its living until the time comes to fatten in the fall. Under such circumstances there is generally a good profit in turkey growing unless losses are very heavy.

The matter of advising turkey culture for farmers or poultrymen located on farms where turkey growing is not common, is one that requires to be handled with caution, and calls for deliberate consideration on the part of those most interested. In some sections where turkeys were once generally grown their culture has been abandoned because of the prevalence of disease among them. In other places, where the natural conditions seem favorable to turkey growing the natural enemies of the turkey are too numerous. Again, where the farms are large enough in area they may be of such proportions that it is impossible to allow a flock of turkeys liberty without their being somewhat of an annoyance to neighbors.

Varieties of Turkeys.

Our domestic turkeys are all supposed to be descendants of the wild turkey still found in its wild state in many places on this continent. The wild turkeys are hunted as game; are often captured, and have been quite extensively crossed with the domestic races to restore the vigor of worn out stocks. In appearance the wild turkey resembles the Bronze turkey more closely than any other tame variety, and by some the Bronze turkey is regarded as simply a domesticated stock of the native wild turkey, modified and improved as is usual when wild stocks of any kind are given the more favorable conditions for development which are found under domestication with good care and liberal feeding.

Bronze Turkeys are the largest, hardiest, and by far the most popular of the varieties. An adult cock sometimes weighs over 50 lbs., though the Standard weights are much lower, i. e., adult cock 36 lbs.; yearling cock 33 lbs.; cockerel 25 lbs.; hen 20 lbs.; pullet 16 lbs. The Bronze turkey is more popular than all other varieties combined, in most sections where turkey culture is extensively carried on.

White Holland Turkeys are second to the Bronze variety in popularity in many sections. They are much smaller, the Standard weights being, cock 26 lbs.; cockerel 16 lbs.; hen 16 lbs.; pullet 10 lbs. Breeders of White Holland turkeys claim that they are much more docile and have less of the roving disposition of the Bronze, and are better adapted to close quarters or a narrow range.

Narragansett Turkeys are often mentioned as the favorite variety with Rhode Island growers, but in a two days trip through the best turkey growing section of that state with a friend a few years ago we found very few of them, both Bronze and White Holland seeming more numerous. We did, however, see a few very fine specimens of the Narragansett. They are second in size, the Standard weights being, cock 30 lbs.; cockerel 20 lbs.; hen 18 lbs.; pullet 12 lbs. In color they are gray, the plumage having a black ground with each feather tipped with a gray band edged with black.

The other three varieties of the turkey are Buff, Slate, and Black. All are quite rare in this country, and good colored specimens of the Buff and Slate varieties are very rare indeed. The weights for these three varieties are the same, cock 27 lbs.; cockerel 18 lbs.; hen 18 lbs.; pullet 12 lbs.

Accommodations for Turkeys.

The turkey is essentially an out door fowl. Like the goose, it will generally by preference remain out doors in all weathers, though that a part of this preference is due to habit, would seem apparent from the fact that in continued very severe weather when they cannot feed out doors, they seem to appreciate shelter and a comfortable spot to feed in. I kept a few turkeys in a room in my stable one season, letting them run about the barnyard and dooryard, but while the old birds seemed contented to keep within bounds, and were thrifty and bred well, they had to be driven into the shed nearly every night, and the young ones, after two weeks old, were anywhere and everywhere but at home.

Bronze Turkeys.

Turkey growers generally provide roosting places for their turkeys out doors in a sheltered place where the roosts can be high enough from the ground to protect them from marauding animals. The sheltered side of a barn or other outbuilding is a favored place for the turkey roost. Occasionally a high open shed is used.

On most farms where turkeys are grown, the breeding stock is either not yarded at all, or yarded only during the breeding season, to prevent the hens straying away and hiding their nests. The hen turkey is very cunning in this matter, and I have known of their going a mile or more from home to lay, going and returning daily, and taking many precautions to throw anyone following off the track.

Some turkey growers yard their turkeys during the laying season. Some yard them until after the eggs have been laid each day, giving them liberty then, and of course seeing that they are back in the yard, and the gates closed, at nightfall. When turkeys are confined only during a part of the day, a yard 50 x 100 ft. will do for a flock of twenty hens or less. The fence, if of wire netting, need not be over 5 ft. high. A few turkeys might be kept permanently in an enclosure no larger than this, but when that is done higher fences are necessary, for immature fowls and hens that are not laying fly much more freely than the mature hens do during the laying period.

The Breeding Stock.

For breeding turkeys, for market extra large birds are not desirable, especially in the Bronze, which sometimes attains such mammoth proportions. A medium sized turkey, plump in body, fine in bone, and active and vigorous, makes the best breeder. For breeding exhibition stock the largest birds are quite generally preferred, size being of great importance in the show room.

Turkeys mature slowly. The males do not reach their full growth until three years; the females not until two years of age. The male under two years old is not considered desirable as a breeder, and the male three years old should be at his best, and good for service for five or six years.

The usual proportion of males to females is one to ten or twelve, but sometimes a much larger number of females is given. A single impregnation fertilizes all the eggs in a litter, and often all laid during an entire season.

Turkey hens are not usually prolific layers. Some lay only a few eggs, perhaps not more than eight or ten before going broody; few will lay more than twice that number during a single period, and eighteen or twenty eggs per hen is about the average for the entire season, though an occasional hen may lay fifty or sixty.

The hens begin laying in March or April. If they are yarded, nests should be distributed about the yard, preferably in somewhat secluded places. Sometimes boxes or barrels are used; sometimes a good sized coop is made. One grower makes his coops three feet square on the ground, two and one-half feet high in front, and a foot and a half high in the back, with board roof and a good sized opening in front for the hen to go in and out. The hen turkey is sometimes very shy about nesting, and will avoid the nests provided for her. To accommodate such hens the breeder sometimes simply puts a bunch of hay or straw in half concealed places, leaving the hen to fix it up to suit herself. Even when the turkeys are given full liberty it is a good plan to place nests for them near the dwelling house, and prevent the hens wandering off until they have chosen one of these nests and laid in it. After having laid once in a nest the turkey, like other fowls, generally returns to it, though some will change if the eggs are taken from the nest. To avoid this, and still keep the turkey eggs from being chilled or broken, some put a few hens' eggs in the nests, and leave them there constantly, the turkey eggs being removed as laid.

One of the most successful turkey growers of my acquaintance, besides putting nests near the buildings to attract the turkeys, uses a coop as shown on next page to cover and protect the nest a turkey makes for herself in the open, and when the turkey, as often happens, makes her nest in an exposed place where the surface does not admit of protecting in this way, builds such a cover over it as is adapted to the location.

Such precautions as these contribute in no small measure to the success of the turkey grower's work, a point which we will have to take up again and again before the end of this lesson.

When the breeding turkeys have full liberty on such a range as a farm adapted to turkey culture usually affords, they will get the most of their living by foraging, and will require in addition only a little grain. Many growers feed corn exclusively, but a mixture of wheat, oats, and corn is generally preferred. Most of the authorities on turkey growing condemn the use of mashes. I cannot say from personal experience just how far their views are influenced by the effects of the use of poor mashes. In my own limited experience in turkey growing, I found both old stock and young throve well on exactly the same mash grain ration I was feeding to hens and chickens.

Turkey's Nest Protected by Coop.

Whether a mash is used or not, yarded turkeys must be fed liberally, and with a good variety of food, grains, vegetable food, meat and bone; shell, grit, and water must be furnished in abundance.

Hatching the Turkeys.

In the rearing of turkeys natural methods are used almost exclusively, and as the hen turkey lays so few eggs it is customary to use chicken hens to hatch and brood the earliest turkeys. The general opinion is that the turkeys reared with hens never do so well as those hatched and reared by turkey mothers. One writer who had had a good deal of experience both ways, said of the relative advantages and disadvantages of chicken and turkey hens as mothers: — "I have found the advantages of turkey hens as mothers as follows: — They are more quiet with little ones; are better protectors from hawks and animals; will not wean their turkeys so soon as chicken hens; are kinder to little turkeys other than their own broods; are better foragers; will take their little ones to the range where they can pick insects, grass seeds, etc.; the little ones are not subject to so many lice as when running with a chicken hen. The main objection to turkey hens is, they are troublesome about coming to the accustomed roosting place with the brood and getting them sheltered for the night.

"Advantages of a chicken hen are that the little turkeys will be more tame as a rule than when mothered by a turkey, and the hen always takes her brood to the coop in the evening and puts them to roost, but as the hen is more restless, she keeps the little ones on the move the first few days, when they ought to be very quiet. This can be overcome by confining her to the coop. I endeavor to set eggs under some chicken hens and some turkey hens at the same time, so when the chicken hen weans her brood they will take up with the turkey hen and her brood and all go together."

When chicken hens are used to hatch turkey eggs the nests are made and the hens during incubation handled just the same as for chickens. The period of incubation for turkey eggs is twenty-eight days, is occasionally as long as thirty days, and the same precautions suggested in regard to the selection of hens to hatch goose eggs should be observed. Nine to eleven turkey eggs are enough for a hen. When turkey hens are used for hatching they may if docile be set wherever the keeper wishes, but if wild must be set on the nests where they have laid. A turkey hen can cover fifteen to twenty eggs.

THE CARE OF YOUNG TURKEYS. 47

Turkey eggs are almost always fertile, and the only test usually made is a few days (three or four) prior to hatching, when if the eggs are put in warm water, those with live chicks in them will "kick." These may then be returned to the nest, and those that give no motion discarded.

The practice when the eggs are hatching varies just as with those hatching chickens. Some advise letting the nests alone until all eggs are hatched; others say, take the little poults from the nest as hatched and put in a warm place, wrapping them in flannel. It is a point each keeper must determine for himself in accordance with the disposition of the hens or turkeys he is using, and also in accordance with his own experience. One person will examine nests without seeming to annoy hens in the least, while another will find his hens resent any interference. This is due to differences in people, probably cannot be changed, and each must act as his experience shows will give him best results.

The Care of Young Turkeys.

No food need be given the young turkeys the first day, but the mother should be fed. If the nest is so situated that a small pen can be made in front of it, it is a good plan to let the brood remain there quietly for some days. If the nest is not suitably located for this, the brood should be removed to a coop with small run attached. A coop with a board bottom is preferred, as the young turkeys are very susceptible to dampness, and a floored coop is dryer.

The pen for a brood of little turkeys should be made of boards twelve to sixteen inches wide, set on edge, and enclosing a space about eight feet wide one way by twelve to sixteen feet the other. The little turkeys are at first much less active and rugged than little chickens, and should be confined to this pen for about ten days. It is better that, if possible, this pen should be on grass land, for little turkeys, like goslings, want green food from the start. If they cannot have a green run green food should be provided. Grit and charcoal must also be provided, and lice must be kept down by dusting with an insecticide the same as with young chickens.

Most authorities insist on the necessity of altogether avoiding dampness, keeping the young

Turkey's Nest in Cleft of Rock, with Loose Boards.

poults up while the grass is wet; but some of the most successful growers I have known let their turkeys range freely after they are strong enough, and say they observe no ill effects from such wettings as the chicks get. Indeed, their opinion is that this way of life is much better than the coddling methods.

While the young turkeys are confined to coops the ground under them and the coop itself should be kept clean. With the ground shifting the coop is all that is necessary if there is opportunity to do that. The coop must be kept dry, as well as clean. In wet weather when the floor becomes damp it should be cleaned daily, and a liberal sprinkling of dry earth, or fine dry litter be applied.

By the time the young turkeys are two weeks old, if not before, the pen will no longer restrain them. Their roving instincts become plainly manifest, and in a very short time after they begin to go over the low sides of the pen they will go over fences four or five feet high with equal ease, and begin to make quite a circuit in foraging. Whether with hen or turkey mothers they require a good deal of watching at this age. In fact while on a good range the turkeys pick a considerable part of their living, and need be little expense for food until fattening time arrives. They require more or less watching throughout the season, and an essential factor in successful turkey culture is to have someone keep an eye on the flock almost constantly to see that they do not wander too far, are not overtaken by violent storms, or picked off one by one by their natural enemies.

Feeding Young Turkeys.

In the poultry literature of a decade ago, instructions to turkey growers make much of the matter of the correct feeding of young turkeys, and each writer was, as a rule, positive that his way was the right way, and the only right way to start young turkeys. This is a point that has come up before in these lessons in our discussions of the feeding of other kinds of fowls, and I presume the reader is therefore anticipating the statement that there are numerous good ways of feeding. A few of these are presented herewith, some dealing briefly with the early feeding only; others giving directions for feeding throughout the season:

Ration I. *For Young Turkeys.* — "After about thirty-six hours old, or after the hen leaves her nest, we feed for three or four meals, equal parts of hard boiled eggs and stale bread. After that mostly stale bread moistened with milk. For two or three weeks we give curdled milk to drink. After two weeks we mix a little red pepper with the bread twice a week." — CRANGLE.

Ration II. *For Young Turkeys.* — "Our first feed is bread and milk, with the milk so pressed out that the bread will crumble. This is fed for the first two weeks, after which the feed is gradually changed to milk curd and meal, one-half part each, and a little cracked corn is given at night." — CURTISS.

Ration III. *For Young Turkeys on Good Range.* — "The first feed I give is milk curd, with onion tops and tongue or pepper grass, cut very fine, seasoned with black pepper. I give this morning, noon, and night. It is a mistake to feed very often, or too much while they are young. If poults are fed three times a day from the time they are hatched until they are grown, they are fed often enough. Yet they must have something to pick all the time, hence I would advise that they be kept in a grass yard where the grass is kept low.

"As they grow older I add other things to the food. Table scraps are splendid for them. If I have infertile incubator eggs I boil them and mix with the other food, but never use fresh eggs, simply because I do not consider it necessary. I give milk instead of drinking water when it is plentiful. I keep grit constantly before them. Wheat is one of the finest feeds for young turkeys. Cracked corn is splendid when they are older. My rule has been to mix grains of wheat in the food from the first, so that when they are old enough to change from curd to grain it will not be so hard to change foods." — Mrs. MACKEY.

Ration IV. *For Turkeys from Shell to Market.* — "I feed poults every two hours until about ten days old, giving stale light bread softened in sweet milk, (or water), squeezed dry, mixed with hard boiled eggs, including shells finely broken. This food is alternated with bread and clabber cheese, oat flake and egg, or cheese seasoned with a little salt and pepper. After the little ones are about a week old I begin mixing a little whole or cracked wheat, Indian corn, Kaffir corn, or millet with the cooked food, and thus they learn to eat grain. Always try to feed no more than they will eat up clean each time. When they are about a week old I begin to drop the white bread, and give them instead what I call a brown light bread made the same as white bread, using one-half white flour, (a cheap grade will do), and the other half about equal parts of shorts and bran, with a handful or two of corn meal. The meal makes it crumble easily. The bread should be allowed to dry for a day or two before feeding; if fed fresh it may choke the poults. I gradually drop the white bread and eggs, and feed instead the brown bread and cheese. When about six weeks old they have become accustomed to the grain food, which, since they were three weeks old, has been kept by them in troughs in coops so constructed that the little turkeys can get in and the older fowls are kept out. By the time the poults are nine or ten weeks old I have dropped the soft or cooked feed to once or twice per day. By September the older poults are dependent upon grain food and range. For fattening I had good results with a mixture of grains proportioned as follows: Two bushels whole corn, two bushels cracked corn, one bushel oats, one bushel Kaffir corn."—Mrs. HARGRAVE.

Ration V. *An All Corn Ration.* — " Successful Rhode Island growers, as a rule, feed their turkeys from start to finish on northern white flint corn, which they grow themselves. They take great pains to feed nothing but well seasoned old corn, because they have found that new corn causes bowel trouble. Turkeys not only like northern flint corn best, and fatten best on it, but it makes their flesh more tender, juicy, and delicious. That given the little ones is coarsely ground, and mixed with sweet or sour milk, or made into bread that is moistened with milk. This is gradually mixed with cracked corn, which, when they are about eight weeks old, is fed clear or mixed with sour milk. In the fall whole corn is given. After June 1st those at full liberty are usually fed but twice daily. They are hunted up and fed in the fields, that they may stay away from the farmyard and out buildings. Many give the turkeys no food from August 1st until cool weather. They get their own living until they come up from the fields in September or October. Upon the approach of cold weather they come to the house to be fed, and thereafter roam but little.

"To fatten them for Thanksgiving, they are fed in November all the whole corn they will eat three times per day. It is not necessary to coop them. The full feeding causes them to rest and sun themselves. Dough is not much used for fattening in Rhode Island. One grower who gives it every morning, and whole corn at night, mixes condition powder with the dough, and finds it causes them to eat more and gain faster. Some raisers give a little new corn mixed with the old at this time, but most consider it safer to feed clear old corn. It is not best to heavily feed turkeys that are to be held for a later market, or those to be kept over for breeding."—CUSHMAN.

Supplementing these rations let me say that in an extended trip among Rhode Island turkey growers, I found it the general opinion of the more successful that with other conditions favorable, the method of feeding was not of great importance, provided the young turkeys got enough to eat, and a variety. These successful turkey growers also agreed that *care* was of first importance, and one woman who was said to have been uniformly successful for many years, made the following interesting observation on the subject of care in relation to feeding; she said that the only difference she could see between her method of handling turkeys and that of various neighbors who were less successful, was that she made it a rule to hunt the turkeys up in the fields on wet or dull days when they did not forage freely, and when insects to be secured by foraging were far less numerous than on bright days, while most of her neighbors fed the turkeys on fine days when food was more abundant and more easily obtained, and sat indoors and left the turkeys to shift for themselves when the weather was bad. When left to themselves with opportunity to range freely turkeys will usually leave their roosting place in the morning and make quite a wide circuit of the fields and meadows, terminating late in the day at the point from which they set out. I well remember how when a boy I used to go occasionally to the farm of a relative who reared every year a flock of 100 to 150 turkeys. There were several boys in the family, all then quite small, whose duty it was to take turns in guarding and herding the turkeys to prevent their scattering and stragglers being lost. It did not require constant attention, but the turkeys were rarely left alone for more than an hour at a time when out of sight of the dwelling. With this constant attention the losses were very small, and at the end of the season the addition to the family income resulting from the sale of the turkeys was quite a substantial amount.

The opportunity for someone whose time is not required for other work to give this intermittent but regular care to the turkeys is almost invariably a feature of successful turkey grow-

Turkey Roost in Shelter of Barn.

ing. Their roving habits make them especially open to attacks of their enemies, and unless they are watched quite closely, all that they represent may be lost in a very short time. For this reason it is generally more satisfactory in the end to grow a flock large enough to make the watching worth while, than to grow a smaller number and let them take their chances.

Turkeys are salable throughout the year, but the best demand and best prices are for "Thanksgiving turkeys." In some of the eastern markets there has grown up within a few years a demand for "turkey broilers," that is, young turkeys weighing three or four pounds each. These are wanted during the summer months, and the prices paid for them are good enough to make many growers prefer to market them then rather than take the risk of carrying them until fall. In eastern localities where turkeys have been long grown on the same ground, losses from disease are often heavy, and the most troublesome diseases develop after the turkeys are a third to a half grown. The grower who anticipates trouble may sell his young turkeys as fast as fit for broilers, but those who expect to carry them through with slight loss generally prefer to hold them for the winter market.

The turkey grower generally sells all that are ready at Thanksgiving. When the fall has been

CATCHING AND KILLING TURKEYS.

favorable for fattening, this takes most of the early turkeys. Then for the Christmas market as many as possible of the later turkeys are slaughtered, and what are left are sold as they become fit.

The dressing of turkeys requires no special description. The methods employed are the same as for chickens, ducks, and geese — dry picking being advisable for all stock to be sold in an eastern market. It is in catching the turkeys to kill that precautions not so necessary with other fowls must be taken to prevent these nervous, shy, and heavy birds from bruising themselves and so damaging the appearance of the dressed carcass, and reducing the price. When the turkeys have been accustomed to roosting in a shed, catching them properly presents no difficulty. When they have roosted in trees or on poles in the open air, catching them requires special care. The method has been described as follows:—

"The usual plan is to get the birds into a barn or carriage shed, and shut them in. In order to do this, they are fed for a long time in front of, or just within the place where they are to be caught. Later, the feed is placed within the building, and they become so familiar with it that they are unsuspicious when within. When they are to be caught, the doors of the building are suddenly closed; or a covered yard of wire netting is built in front of the building and closed when all are in. Usually when they find they are confined they become frightened, and fly back and forth, or huddle up in corners. * * * To overcome this drawback, certain raisers have improved the usual makeshift catching place by building a long, low, dark pen back of the barn or shed. This pen extends alongside of the building, and is at right angles with the entrance to it, and at the extreme end is about two feet high. Up to the time of their being caught, the end is left open, and the birds frequently find their way through it. When they are to be caught, only what the pen will comfortably take are driven in. They do not discover that the end is closed until it is too late to turn back. The turkeys that are not to be caught, are first driven away; otherwise they may be alarmed, and become unmanageable. No turkey that is thus caught, and has learned the mysteries of the trap is ever allowed to escape, or its suspicions would be communicated to the others. When shut in this pen they are quiet, and when a man goes to catch them, there is no struggle; he simply reaches out and takes them by the legs. The pen is too dark and narrow for them to fly, and too low for them to crowd one upon another."

LESSON VI.

Possibilities and Probabilities in Poultry Culture.

IN THIS lesson I am going to depart a little from the usual plan of the lessons. There are several reasons for doing this. First of all — the subject is too big for a lesson of the usual length. Next, I think it can be presented most effectively by treating it in three principal divisions. And, finally, I think that the points it makes in the concluding division will be much more generally appreciated if illustrative examples are used much more freely than has been found practicable in the lessons in general.

So we take the subject up in three sections. The first section will consist of a reprint of a part of a lecture given by me at various places some years ago, and published in full in the issue of this paper for August 15, 1902. The second section will be in narrative form, and will tell of the poultry experiences of a large number of persons—both those who have failed, and those who have succeeded in poultry keeping. The third section will give in systematic form a statement of conditions and circumstances making the possibilities, or affecting the probabilities of success in any line of poultry keeping.

I hope to be able in this way so to present the subject that anyone can determine for himself what is best for him to do, and what lines it will be most advantageous for him to follow.

Why the Failures in Poultry Keeping.

It is commonly asserted that over ninety per cent of all business ventures fail. On what authority this assertion is based I have never been able to learn, but I have very strong doubts of its accuracy. It is also commonly believed that the proportion of failures to successes is very much larger in the poultry business than in almost any other line.

If this is so, and if the general statement in regard to business failures is correct, the percentage of successes in poultry keeping would have to be very small indeed.

It is not now, and perhaps never will be, possible to get accurate data on this point, but we can still make comparisons which will have some value in indicating the relative number of failures among poultry keepers. Some months ago, just to satisfy my own curiosity, I took a copy of a poultry paper for March, 1891 — ten years ago — and went through the classified advertisements in it, checking the names of advertisers who, to my knowledge, were still engaged in the poultry business, and including with these the names of a few who had died while active in the business. Of 240 persons advertising stock in that paper of ten years ago, I found that 50 — more than twenty per cent — were still engaged in the business. As I omitted the names of a number I think (but do not positively know) are still in the business, and as it is probable that a number of the others of whom I know nothing at all are still in the business, perhaps a thorough investigation would show nearly as many more still interested in poultry keeping. So that that method of getting an indication of the proportion of failures would indicate that they were not more than sixty to sixty-five per cent.

WHY THE FAILURES IN POULTRY KEEPING.

But with the best showing that could possibly be made in this way, it would still remain true that a majority of those who undertake poultry keeping — whether for profit or pleasure — fail to realize on their expectations; and as we can only judge what is to be by what has been — making due allowance for general progress — we have to consider it as improbable that the proportion of failures to successes will be much reduced in the immediate future.

Then putting the matter in its most favorable aspect we have to say that more than half of those who engage in poultry keeping will make a failure of it. In other words, that when one undertakes poultry keeping the chances are against his success.

If every prospective poultry keeper could be made to appreciate this before he begins, the proportion of failures might be so much reduced that successes would preponderate; but, unfortunately, nearly every beginner thinks himself, or herself, the talented and favored or exceptionally industrious person who is sure to succeed, and therefore neglects to take the necessary precautions to avoid failures, even when advised of them.

With very rare exceptions, those engaging in poultry keeping on any considerable scale begin without any adequate practical knowledge of the conditions, requirements, and methods of the business. Sometimes they have had a limited experience with a few fowls; but quite as often they have had no actual experience, and a very limited and superficial information. If they happen to have abundant capital to carry on the business until they have learned in the costly school of their own experience what they ought to have known before investing a

View of Grandview Poultry Yards, Aurora, N. Y.

dollar, they may make a success of it — may finally make the business pay its expenses, and give them a living besides; but the greater part of the original investment, and also of the expenses for several years, may have to be charged off to cost of acquiring experience — that is, cost of learning the business.

But the most serious cases are those of people with limited means who go into poultry keeping expecting it to yield an income sufficient to meet all expenses and give them their living almost from the start. These people engage in the business with two ideas which I think are either separately or jointly responsible for their undertaking it — both of which ideas are radically wrong.

The first of these ideas is :—that poultry keeping is very easy, that there is really nothing to learn which any person of average intelligence cannot acquire at once and almost without effort as soon as ever he gives his attention to the subject, and that the fowls require so little care that their owner has light work and a great deal of leisure.

The second idea is that the profits of the business are very large, and the margin of profit in each of the branches of the business so good that even if there are losses due to inexperience these cannot possibly be heavy enough to make the business run behind.

Novices are not wholly to blame for these errors unless they persist in them after the facts have been clearly presented. The people most to blame are the interested parties who circulate such ideas. But, as far as my observation goes, the greater number of persons who once

Feeding, Watering and Egg Collecting Cart Used on a Rhode Island Farm.

become deeply interested in poultry, and decide to venture into it will take bad advice in preference to good every time. I suppose this is because the bad advice is more in line with their hopes and wishes, and because those who give them good advice will admit that though the chances are very much against the success of a business established on the basis they propose, there is still a chance that intelligent application and hard work will pull them through.

The daily care of poultry is neither as easy as some think it, nor as hard as some others make it. It is easy when you know how, and, unless you happen to be one of those who utterly lack natural aptitude for handling live stock, you find it neither a long nor a difficult task to *know how*. But whoever without previous experience undertakes the care of a large stock of poultry, soon finds himself in the predicament of every man who undertakes to do or to learn too many different things at once.

There is a very great difference between doing work well and doing it profitably. Permanent success in any line of work depends, as a rule, upon doing it both well and profitably. The workman must combine thoroughness with a considerable degree of speed,—he must have skill and facility. Skill and facility come only as the result of thorough practice so long continued

Chickens in Corn Fields, on Farm of Knapp & Son, Fabius, N. Y.

that a piece or kind of work becomes largely mechanical, head and hand working together without conscious effort to keep their operations in harmony.

The novice in poultry keeping who undertakes to establish and manage a plant all by himself, has a great variety of unfamiliar things to think about. As the Irishman expressed it in his parody on Poe's well known poem, "The Raven," he has to spend a great deal of time and gray matter in "thinkin' thoughts he never thought of thinkin' of before." He also has to learn to do a number of things quite new to him, and as a result he does a great deal of hard work both mental and physical, and has very little to show for it.

Frequently he has never even seen a well equipped poultry plant, and has no knowledge at all of good methods of doing the work on his plant. Often he makes very hard work of very simple things just because he has no one to tell him or show him what to do.

But, after all, the work of caring for the poultry and the worry which this work causes an inexperienced man or woman constitute the least part of the inexperienced proprietor's real troubles. Even before this work begins he is called upon to decide matters of deepest importance to the success of his business while, as yet, his opinions on the subjects involved are not worth a cent to himself or to anyone else. The result of this is that such beginners are constantly doing things which to persons having any real working knowledge of the business, and likewise to all persons in the habit of exercising plain practical common sense appear incredibly stupid and altogether inexcusable.

I had some correspondence not long ago with a man who was having a great deal of trouble with roup and rheumatism in his flock. He used the most approved remedies for both without securing any permanent improvement in the general condition of his flock. After repeated inquiries as to conditions I learned that his houses were located on low damp ground, where both soil and atmosphere were objectionable from a poultryman's points of view. His houses were only dry in periods of drouth, and often there was stagnant water all around them. He had not even the excuse of having had the place already on his hands before he went into poultry to offer, nor could he plead that he did not know that the location was unfavorable.

He had bought this land for a poultry farm because he had made up his mind to try poultry keeping, and this was the only farm he had been able to find within the limited time he allowed himself to find a farm that came within his means, and he had thought that even with the disadvantages of this location he could — by giving his fowls special care — make enough in a few years to buy a more suitable farm.

I could give numerous similar instances where people have deliberately gone contrary to the known teachings of experience, because that experience not being personal to themselves, they could not realize the danger and folly of rejecting its lessons. Had they learned the business first under a competent instructor they would not have been likely to go badly wrong, for the habit of doing a thing right often keeps one out of trouble, even if he does not understand the reasons for the method he uses — and has never seen a practical demonstration of the consequences of doing that particular thing in some wrong way. Someone has said, "Success does not consist in never making mistakes, but in never making the same mistake twice." I don't think that assertion will bear a very close analysis; very few generalizations of the kind will; but it certainly has a great deal of truth in it.

As I look back over the years when I was learning poultry keeping—experience made largely of mistakes—I recall that at the close of each season I used to note, in reviewing that season's work, that my marked progress had been principally along the one or two lines in which I had found most discouragement and loss in the previous season, and to which I had therefore given most thought and attention; and in the next season work on those lines was comparatively easy, had perhaps advanced to the mechanical stage, and more time could be given to some other troublesome matter.

This was progress, and there was a certain satisfaction in working things out for oneself, but it was a slow and laborious progress, and the cost was enormously greater than if I had learned the business in the right way.

There is another way in which poultry keeping is hard for most people—and very hard for some—which few think of until they learn it by experience.

The business is very confining, and so in time becomes monotonous.

Then it becomes a question of whether the poultry keeper can take up some form of recreation that can be adapted to such leisure as he can secure, and furnish relaxation which will break the monotony of his work without interfering with it. If he cannot do this—and if he is of such a disposition that he cannot stand the monotony of the life, he is very apt to begin to cut his duty here and there to get time for favorite pleasures, and when he does this the finish of his venture is only a question of time.

The poultry keeper—like all who have the care of live stock—has to give the real needs of his stock precedence over all ordinary claims upon his time and attention. The care of the stock will frequently require long days of labor extended far into the night, loss of sleep, and denial of many pleasures.

"Business first" must be his invariable rule, for there is no other line of work in which the penalties of slight infringements of that rule are more sure or more quickly felt. Whoever finds it too hard to follow that rule will fail in poultry keeping.

A common cause of failure—which is in part the cause of by far the greater number of failures in poultry keeping—is a lack of sufficient capital. Any business undertaken with insufficient capital is heavily handicapped at the start. In poultry keeping it is almost the rule for men to begin with an amount of available capital which is insignificant in proportion to the amount actually required by their plans.

Those supposed authorities on poultry keeping who have so industriously preached that the poultry business requires smaller capital and will yield larger returns on the investment than any other, have a great many failures to answer for, and so have those who have advised prospective poultrymen to go ahead on a capital which they could not help knowing was insufficient.

While I blame such persons for misleading people, I have not as much sympathy as some have for those who allow themselves to be misled, and have not much patience with them when they try to put all the responsibility for their failures on those who advised them badly, because too many such cases come to my notice where people have also been given good advice —but have followed the bad because it was more in accordance with their desires.

I sometimes think that most of the people who ask advice about going into the poultry business are simply looking for encouragement to go ahead with plans—which are often very peculiar plans,—and keep asking until they get the kind of advice or encouragement they want. In such cases as this, both adviser and advised are equally responsible for the failure, but the division of responsibility does not diminish the share of either.

Far too many of those who build poultry plants have not capital enough to properly equip and stock them—to say nothing of running them until the profits begin to appear. And so financial worry is added to all the other worries. The poultryman goes into debt—practically mortgages his receipts for months in advance—carries on his business in a hand to mouth way, neither buying nor selling to best advantage; interest eats up a large part of his profits, and finally he is forced to the wall.

Lack of business ability is responsible for many failures.

It is very difficult—if not quite impossible—for one who is not a fairly good business man to make much of a success of poultry keeping, and in some branches of the business a man is seriously handicapped if he is not a good correspondent and salesman.

My observation has been that the poultrymen who lack business ability—who are deficient in the trading faculty, seldom realize that the fault is with themselves. Many of them are disposed to quarrel with the conditions of the business and to imagine all kinds of crookedness and meanness the most substantial aids to the advancement of their more successful competitors.

I suppose that in New England it is not necessary to dwell long on the lack of business ability, for Yankees are supposed to be born traders, and if a genuine Yankee fails in poultry keeping we have to lay it to one of the other causes.

The three things named — inexperience, lack of capital, and lack of business ability, I consider the principal causes of failures in poultry keeping; but, besides these there are numerous minor causes which frequently prevent success, or turn most promising prospects into failure.

CRUDE IDEAS AND POOR METHODS.

Perhaps in a strictly systematic treatment of the subject some of the minor matters should be classified under the others and considered as subordinate to them; but I do not think that to do so would give them any more force, and it seems to me much the better way to consider them independently.

A number of errors similar to that of selecting a bad location, which has been mentioned, are common. Sometimes these errors are the results of inexperience, sometimes they are due to the prejudices or follies of poultrymen of experience enough to have acquired good judgment in the matters involved. Of this kind are errors in poultry house construction—not the minor errors, but the big mistakes — the mistakes that are so absurd that they are serious. There are a great many poultrymen — and not all of them inexperienced novices — who, when they get an idea which they think would work well in a poultry house, are not satisfied to test it on a small scale first, but must apply it, at whatever cost, to one large building at least — if not to the whole plant.

Breeding House at Jordan Farm, Hingham, Mass.

There are hundreds of poultry houses in this country where the incorporation of a few "original ideas" is costing a great deal in extra work and wasted time; hundreds where wrong construction makes it unnecessarily hard to keep fowls healthy and productive. There is absolutely no excuse for this, for the construction of a good poultry house is about as simple in theory as the construction of a good dry goods box, and it is hardly more difficult in practice.

Too many people seem to think that changing a good plan is improving it, especially if they think the idea of the change original with themselves; and such people are not apt to accept the testimony of others as to results of putting their ideas into practice. As a rule the miscarriage of their plan will convince them that it is wrong.

In justice to the general good sense of poultrymen, however, it ought to be said that the majority of them are quick to see errors of this kind when use brings them out, and would be quick to correct them if they could take time to do so, or could stand the expense.

A great many poultrymen lose time and money by clinging to poor methods of doing work. Indeed, almost all poultrymen lose in this way. That is one of the disadvantages of being mostly self taught in anything. One works out a poor method, and after that becomes a habit finds it hard to change. I have known large poultry farms developed from very small beginnings where methods which were all very well for a few dozen hens and chickens, but wholly inadequate to doing the work economically for a few hundreds, were continued when the number of fowls and chicks aggregated several thousands.

If this fault occurs only at one or two points it may not make the larger business a failure — though it will surely cut the profits; but if it is general it is sure to make a failure, and it is because they do not develop methods suitable to their increased stock that so many poultrymen who are successful on a small scale fail to do well on a large scale.

After a poultry plant is once built and stocked the most important item of expense is the

Capons in Colony Houses at Jordan Farm, Hingham, Mass.

labor. The cost of feed may be greater, but the cost of labor is more important because more difficult to regulate.

A well known poultryman once said — referring to someone else's habit of getting up early and having a great part of the routine morning work on his plant out of the way before breakfast — that there was no need of getting up so early if one would lie awake long enough at night to plan easy ways of doing his work.

I don't think he meant this to be taken literally. I imagine his idea was to convey as forcibly as possible the idea that to learn to work well — particularly where there are many different tasks to be fitted into a day's work — one must do a great deal of thinking and planning for the work.

A half an hour is not very much time; but see what a saving of time a poultryman would effect who would so rearrange his work, or so improve some method that he would save a half an hour a day. Most poultrymen work 365 days in the year. A half hour saved each day would mean in the aggregate three weeks of six days of ten hours each — an item worth looking after. This time, properly used, would enable many a poultryman to do a great many

LACK OF JUDGMENT AND STABILITY.

things left undone for lack of time — there are plenty of such things on the average poultry plant — or take a little needed recreation.

There are a great many old poultrymen who have their work so well systematized that it would be hard to plan such a saving in time as this. The old hand's shortcomings in such matters are generally limited to occasional tasks. His regular work as a rule he has, as the saying is, down fine. But nearly all beginners, and most of those who, after a few years hard work, are still creeping along on the ragged edge of failure, could save much more than a half an hour each day.

The old method of learning a trade, when a boy worked as an apprentice for a number of years, and then as a journeyman traveled about, working a short time in each of a number of places, is the ideal way of learning poultry keeping. I sometimes think that we will never know just what can be done with poultry until we have among poultrymen a considerable number of bright intelligent men who have grown up in the business, and thus learned it more thoroughly than most of those who pick it up later in life ever can know it. When that time comes we may look to see successful poultry plants on a very large scale — and until then I do not think we shall.

I have already referred incidentally to the aptitude for the work of caring for live stock as a factor in successful poultry keeping. If we attempt to analyze this faculty we find that it consists mostly of good judgment as to the condition and needs of each animal; and if we try to learn the history, or trace the development of this faculty in individuals, we find that it is a natural talent developed by experience and training. Where the talent is conspicuous, the person possessing it will be quite successful from the first with almost any kind of live stock, and in time will become notably successful. Where it is of less degree, experience and training — in inverse ratio to the amount of talent — are required to make one proficient in the management of live stock. Where this faculty or talent is wholly wanting, I do not think it possible for the person so deficient to ever attain any respectable measure of success. His occasional successes will, as a rule, be purely accidental. The proportion of persons thus deficient is probably small — or appears so because few of them attempt to go into stock breeding. Yet, first and last, a great many such persons do engage in poultry keeping, and if they are of persevering disposition, peg away at it for a long time before they come to a realization of their unfitness for the work.

A very common idea, which seems to me wholly wrong, is that love of animals is an important factor in ability to manage them well.

A great many prospective poultry keepers mention that as the first and most important of their qualifications for making a success with poultry.

As far as I have been able to analyze the conditions of success in handling live stock, a love, or strong liking for animals, and good judgment in caring for them are two entirely independent attributes. They are frequently found existing together, and sometimes one helps the other; but an excess of affection for animals is apt to bias one's judgment as to their needs. The cultivation of the calculating spirit in considering animals is quite essential in one who keeps them for profit, and this spirit is likely to develop a very cold blooded matter-of-fact brand of the article called love.

A strain of fickleness in a person's character is likely to develop in various ways when he engages in poultry keeping, and nearly always in ways detrimental to the success of his business in poultry.

Perhaps the most noteworthy illustration of this is found in the case of persons who are continually changing breeds of fowls, never keeping any one long enough to know what it is or what they can do with it. It takes several years of careful handling and close observation to show one just what his stock is, and if he is breeding for fancy points generally several years more are required to get the stock on such footing that he is at all sure of results from it.

This being the case, it is as clearly impossible for one who changes breeds every year or two to make any perceptible progress as it was for the frog in the well which, in the catch problem in the old mental arithmetics, was said to crawl up three feet every day, and slip back four feet every night.

Just at present a desire to be progressive and scientific is hurting a great many novices in poultry keeping.

Poultry and agricultural papers have published numerous articles about the science of poultry feeding, written by persons who knew very little of either science or feeding, and the general impression these have given people who have not learned feeding by practice is that the ration can be, and should be, compounded according to a mathematical formula, and exactly balanced to suit the needs of the fowl — or the requirements of the keeper — the two being assumed to be synonymous.

Now the scientific investigation of articles of poultry food, and of results of methods of poultry feeding is a necessary work which will, in time, no doubt arrive at some important results; but the folly of persons who have not the judgment and skill to take a tested ration, recommended by experienced feeders after years of use, and keep a stock of fowls healthy and productive on that ration, trying to compound rations by formula, and feed them by weight is something truly appalling; and the number of those who are frittering away their time and money while trying to make figures and weights do work for which nature, if she intended them for poultry keepers gave them brains and eyes, is very much greater than is commonly supposed.

Unless restrained within very narrow limits, the disposition to experiment and investigate may prevent one from making poultry keeping financially successful.

Experiments are expensive, and comparatively few of them yield any immediately useful practical results. If carefully and thoroughly followed out, experiments invariably take more of one's time and thought than he intended they should, and it all comes out of time and energy which, if used for work of which the results were practically assured in advance, would help instead of hindering success.

To some, experimental work is in the line of recreation, and in this way it is all very well if not allowed to interfere with regular work more than a recreation should; but, on the whole, and as a general rule, it is much better not to engage in it until one's poultry business is on an assured basis. Even then it must be indulged in but moderately by those keeping poultry for profit. Let them leave it to those who keep poultry for pleasure, and to the experiment stations.

Many who try to make the breeding of fancy or exhibition poultry profitable, fail because they are not, and never can be, fanciers.

To be a successful fancier one must be something of an artist, with a keen appreciation of the points that go to make the ideal fowl in his variety. The artistic faculty is generally a birthright. As the saying goes: "Fanciers are born, not made." Their talent improves with use, but, if small, cannot be developed by training to the same extent that a moderate aptitude for the care of stock may be.

It has seemed to me very noticeable among poultrymen of my acquaintance that the best fanciers have seemed to show from their first acquaintance with a breed or variety a correct appreciation of what a fowl of that breed or variety should be, even though it might be some years before they learned how to produce the desired types from their matings; and it has seemed just as noticeable that others, after years of studied and persistent effort, were as far away from producing what was commonly desired in fowls of the breed they handled as they were at the beginning. In some this was evidently due to inability to appreciate all the details of beauty in the best representatives of the breed. In others it was as clearly due to the absence of a disposition to harmonize their personal tastes with accepted ideals or standards.

The successful fancier must not only have good artistic perceptions, a good eye for form and color, but his practical success depends upon his being conventional, upon being one of those whose ideas naturally harmonize with ideas prevailing about them.

So — it is possible for anyone to make practical test of whether his artistic perceptions are of the degree and quality necessary to make him successful as a breeder of fancy poultry. To make such test he has only to compare his judgment of his birds with that of others, especially of good breeders and competent judges. He will find these agreeing in the main, though often differing in particulars.

LESSON VI.

SECTION II.

Some Typical Ventures in Poultry Keeping.

HERE is one of the most remarkable successes in poultry keeping. Less than twenty years ago a laboring man of foreign birth, living in New England, bought a worn out farm, paying a small amount of cash, and giving a mortgage for the balance. In all his life this man had never earned over a dollar and a half a day. He was industrious and frugal. His wife was the worthy helpmeet of such a man, doing her share to make his slender earnings go as far as possible. His children were trained in the industrious and thrifty habits of their parents, and at the same time given good common school educations.

At the time this farm was bought the sons of the family were young men, working generally, as I have been told, by the day for farmers and others in their vicinity. After the farm was bought the father and his several sons continued day's works for others, carrying on the operations on their own farm at such odd times as their services were not in demand elsewhere, and in this way turning all their spare time to good account.

In their efforts to make the farm profitable they tried various special lines, among them poultry keeping. They established quite a flock of practical thoroughbred fowls, and found them increasingly profitable, and were gradually increasing their equipment and stock, when in some way their attention was turned to duck growing. Just how this came about is not, so far as I have learned, a matter of record; but in view of the fact that this was just at the beginning of the great boom in market duck culture there is nothing at all strange about it. And then there was located not many miles from them, the farm of one of the pioneer duck growers in this country, a man who was doing a large and growing business in market ducks. Had the interest in duck culture been less widespread than it was they could hardly have failed to hear of this man and to be interested in what he was doing. They visited him and decided to begin duck growing in a small way. Within a few years their business had so increased that the entire family were giving all their time to it. The farm had been paid for, a plant representing very much more than the value of the land had been erected. (Though a large plant, it was by no means an elaborate one). Their joint incomes were steadily increasing; they had ample capital to meet the growing demands of their growing business, and secure correspondingly larger returns on the amount invested; the farm acreage was gradually made more productive; new houses were erected. In less than ten years from the start on that farm their business ranked among the first poultry businesses in this country, both in volume and profitableness.

About the same time that operations began on this farm some members of an enormously wealthy family—men, too, who had made their money, and are still making it in their regular business,—established a poultry farm only a few hours distant from that I have just briefly described. They had unlimited land and unlimited capital. They hired a manager at a good salary and built an expensive plant. They went into it for profit, not for fun; but when I saw the plant about ten years after its opening, I was told by one who knew its financial condition intimately, that the plant represented a net loss to its owners of over $20,000. They were men who could stand even heavier loss without financial or personal inconvenience, but they were not willing to run the plant at a loss. That was not what they operated it for. They

House for " Winter Chickens" on a " South Shore " Farm, H. D. Smith, Norwell, Mass.

hired another manager, who was able to make a better financial showing,—yet not, as I judge from the fact that he remained but a few years,—good enough to be encouraging either to him or to the proprietors. The plant represents too much dead capital, and its manager, however competent, would always find himself handicapped in various ways. I am not able to say in just what condition the plant now is, but there is little likelihood of its ever being made profitable enough to wipe out the heavy deficits of its earlier years and leave its owners square.

A good many years ago there were two brothers, young men and unmarried, who jointly bought a little farm away back in the hills in central New York. They had but little money to pay down. Together they worked the farm in summer. In winter one ran the farm and the other taught the district school, assisting with the chores mornings and evenings. They went on this way for several years, making a bare living and just keeping up the interest on the mortgage on the farm.

As they worked together they used to discuss various methods of making the farm more profitable. Their attention was finally drawn to the possibilities of profit in poultry, and after talking it over they concluded — to go into the poultry business? — Oh, no; — to keep strict account for the little flock of fowls they had on the farm, and see for themselves what they could make from a small number.

The flock consisted of less than a score of ordinary fowls. The profit on this flock was so satisfactory that they increased the flock,—about doubled it. That was still a small flock, not at all up in numbers to the ideas of the average beginner of the number with which it is worth while to make a start. The third year they increased in about the same proportion, the flock being still below the hundred mark. After that the same rate of increase made large additions to the flock every year. They began to get into thoroughbred stock; went to local shows and won prizes; went to New York and won more prizes, and began to sell eggs for hatching and exhibition and breeding stock at high prices. Money began to be easier with them. They left the small farm and bought a larger and better one more conveniently located. The profits on poultry gave them the means to enlarge other farm lines.

SOME STRIKING CONTRASTS. 63

Their "fancy" poultry business was but an accessory of their work in practical lines. As soon as they began to have eggs to sell in quantities, one of them went to New York and looked up special customers who would pay extra prices for a good article. Finding the demand too great to be filled by their own supply, they began to collect eggs from their neighbors, and gradually extended the circle of collections until, when I visited the farm last, they were handling about $25,000 of eggs a year. The farm is one of the finest in that section. One of the brothers retired from the firm a few years ago, and the one who remained and his son continue the farm along the same lines. They are also interested in many outside enterprises. The head of the firm said to me a few years ago: "Our poultry gave us our start. We have made more money since from other things than we ever did from poultry; but poultry has always paid us well."

A few years ago a stranger stepped up to me at a New England poultry show, and said: "You were pointed out to me as the editor of FARM-POULTRY. I've just won a lot of prizes here, and I want to advertise in your next paper. I want a big space, and I want your best terms. I've spent a lot of good money for birds to show, and now I want to do business. I believe the way to get business is to sling printers' ink. I've got plenty of money, and I'm going to sling it." By "it" he referred, of course, to the ink first, but indirectly, I suppose, to his money. He had bought a fine farm. He put up expensive buildings. He took in a poultry fancier of more experience as partner. He remained in the business less than two years.

A man engaged in a manufacturing business bought a small farm inside the city limits for a home. There was room for a garden, cow, and of course some poultry. He had never taken any special interest in poultry, but fowls were a necessary part of the equipment of such a place, so he bought a few — just to keep the family in eggs.

Having the fowls he felt that he must know something about fowls—so he began to read a poultry paper. It happened to be FARM-POULTRY. Reading it he became interested in several features in poultry culture. He built a broiler plant, and made something of a reputation in broiler raising. At the same time he began to breed one of the most popular varieties of fowls. He exhibited, and won prizes. He advertised in a small way at first, gradually increasing as his stock and trade warranted the expense, until in a few years he had one of the largest poultry businesses in the country, and the broiler adjunct was dropped because it was found more profitable to devote all attention to the production and sale of high class stock.

About the same time an elderly gentleman of considerable means, bought a farm with the idea of fitting up a poultry plant for his only son who was an invalid and needed outdoor occupation. He spared no expense in equipping it. He employed an "expert" to lay out the

Poultry House in Corner of a Farm Door Yard.

plant and paid roundly for "expert" advice at every turn. He soon found that in order to get anything at all out of the plant he must give it his personal attention, and the farm was kept going for several years on this basis. Meantime the son died, and the father, to whom the plant was now of no use, was glad to sell it for a small part of its cost.

Perhaps thirty years ago the son of a New England farmer, arriving at an age when enterprising young men begin to plan very seriously their life work, concluded to try what he could do with poultry. He began on his father's farm, and with a cash loan to give him a start. He worked for his father for his board, and to repay the loan, while getting his stock established. Beginning as a practical poultryman, he became interested in exhibition stock, was very successful in breeding it, was an excellent salesman, and in a few years built up a business ranking among the largest in the country. As his trade in fine poultry grew he dropped the market side, and gave all attention to the lines that were giving him greatest profit. He has prospered in business. A few years ago he said to a group of friends to whom he had just

White Leghorn Chicks on Farm of H. J. Blanchard, Groton, N. Y.

shown a business block he had recently completed in a town near his home, "This is my provision for my family in case I am taken away. This building will give them a comfortable income." Years ago I heard him say that while he had found poultry keeping profitable, he believed that any man who could make money at poultry could make more money at something else.

Eight or ten years ago a traveling man bought a farm in New Jersey, and put a good sized poultry plant on it. Immediately marvelous tales of its success began to be circulated. His detailed statements of results showed how easy it was to make money with poultry if only you had the personal equipment which everyone who thinks of starting supposes he has. In a very short time the remarkable success of the plant on the established scale indicated such great profits from larger operations that he easily interested capital in his schemes, and the farm was made one of the show farms of America. People came from far and near to see it. The poultry press generally gave it extended write-ups. Then all at once the promoter disappeared, and those who had furnished the capital put the concern into bankruptcy.

HOW ONE MAN'S PLANT GREW.

Not many miles from Boston there is a small farm which the present owner purchased about a dozen or less years ago. He had money enough to pay for the farm, make the old house habitable, buy a small flock of hens, and have a few dollars left for an emergency. For a few years he continued working at his occupation. He saved what he could. The few hens earned something; when he had money enough ahead to buy material for a house he bought it. Then at such odd times as he could he put up the house, at the same time planning to have increased his stock so that when the house was ready he had extra pullets to fill it. He planted fruit trees, and seeded down such parts of the farm as were suitable to grass. He kept a cow or two. Everything was made to contribute something to the total income. It did not take as long to get money ahead for the second house as for the first. When the money was ready the material was bought, and the house built at his convenience. By this time his farm was taking more of his time, the outside work was gradually reduced as home demands became more imperative, and soon his farm was taking all of his time, and he was making a living from it.

A Town Lot Poultry Plant at Wellesley Hills, Mass.

He has now house capacity for 800 to 900 hens, keeps several cows, and makes a comfortable living and a little more. You need not drive far from his place in any direction to see the wrecks of poultry ventures embarked much more auspiciously than his. He is in no sense of the word a fancier. He is even indifferent to thoroughbred stock, using mostly good grades, but his poultry pays.

In that section of the old Bay State celebrated among poultrymen as headquarters for fine market poultry, two brothers began some fifteen or sixteen years ago to try to grow winter chickens. One, who was employed in a factory, had a little money saved up. He bought some incubators, contracted for some eggs, and the other brother went to work to see what he could do with them. Hatches were discouraging. Hundreds of eggs went through the machines without giving any substantial supply of chicks. The greater part of a hundred dollars had been expended in eggs, and the man who was running the machines wanted to quit. He thought it was no use. But the capitalist of the pair insisted on sticking to it as long as he had money or credit to buy eggs, and soon luck changed. They got enough out of the first season's work to encourage them to go ahead. Slowly but surely they did go ahead until they had a plant to which both gave their full time during most of the year, (their special line giving them a month or two off in the summer if they chose to take it), and were making more clear money for each every year than the average professional or business man makes.

Two young men, brothers-in-law, employed in a factory, together built up a poultry business to the stage where one of them, whose health was not good, and to whom constant indoor work was injurious, could give all the time that he was able to work to it, the other helping him out in emergencies. The plant was at the home of the invalid proprietor. They would have continued to develop it until large enough to make a living for both, but there was not land enough, nor was it possible to buy adjoining land. At length an opportunity came to buy a piece of woodland a short distance away. It could be bought for $150. I do not remember the acreage, but it was not large, only enough for a small farm. It was purchased by the second man. He cleared it, and made enough on the sale of the wood to pay for the place, so that he started with the land clear, and what small capital he had of his own could be put into a dwelling, poultry buildings, and equipment. The house erected was small and plain, only sufficient for the actual needs of the family, costing probably not to exceed $500. The outbuildings also were built as economically as possible.

The two men, though not now in partnership, worked on a coöperative plan. The original plant had been run as an egg farm. It had a capacity of about nine hundred hens. By the arrangement made, the owner of this plant furnished the other at market price all the eggs he wanted for incubation, carrying the account until fall, when he took his pay in pullets, at market prices for poultry at the time they were delivered. As the man who grew the chickens hatched both winter and summer chickens, and grew many more than required to furnish the other what pullets he needed, the value of the eggs he took would each year run close to the value of the pullets he delivered. The plan worked very satisfactorily until with the continued ill health of the invalid poultryman, a change of climate became necessary for him. Both farms were sold, and both families moved to a milder climate.

On a Maine farm devoted to general farming I found a flock of between 400 and 500 hens kept in houses built to accommodate 50 to 100 fowls each, these houses being distributed within a radius of the dwelling which made it not too hard a task for the farmer's wife and mother to attend to them during the summer when the men were engaged in the fields, the men taking care of the fowls at other seasons, and also in rough weather. There were too many hens close to the house to admit of keeping the place as we like to see the surroundings of a dwelling; but the farmer said his hens were the best paying stock on the place, and as his method of handling them was adapted to his situation and circumstances, and they were thrifty and productive, he did not feel disposed to make any sacrifice of profit to appearances at present. This flock had been built up very slowly. He had been seven or eight years in getting to the number he had when I saw him. He said the usual increase had been about fifty hens a year, as he had found that he could add that number each year and make the necessary provision for them without taking more of his income for the purpose than he conveniently could.

On many farms in Rhode Island hens are kept by the "colony system." I presume that on most of these farms the beginnings of the system date back for over a generation, and on many much further back than that. I spent parts of two days at different times, going about as he cared for his poultry with a young man whose father and grandfather before him had for years kept poultry on this same farm, by the same methods and with the same kinds of stock. Without going into a full account of these methods here, I will say that with such a small poultry house as might be used anywhere — on farm or town lot — as a unit, anyone who has room to spread his fowls out, locating houses far enough apart so that the flocks will mingle little, can apply this system. The limits of it come with the limits of his land. All that he has to do is put a small new house in a suitable place whenever he is ready to do so. He may use two or three houses, or he may use a hundred; the system is the same, and the conditions the same for all.

HOW TWO WOMEN WENT INTO POULTRY CULTURE.

A school boy became interested in poultry, bred and studied a popular variety, and before he was out of his teens was competing successfully with veteran fanciers. He was educated for a profession, but his health being poor after having embarked in his profession, he found the confinement too much, and engaged temporarily as manager of a poultry plant. The position proving congenial and profitable, he continued in it. Whether he is as well off financially as he might have been had he continued in the line of work first chosen, I cannot say, but he earns a very fair salary now in a position which still leaves him a little time for judging shows, contributing to the poultry press, and mating fowls for others.

A lady who had for years taught school, found after her marriage that housework was far from congenial. After many consultations with the gentleman most interested, she arranged to hire her household work done and give her own attention, except for the cares devolving upon the wife and mother, to poultry, of which she was very fond. Her poultry business grew and flourished until the problem became how to keep it from intruding upon other obligations, and at last was given up solely because it was found impossible to make a satisfactory adjustment of business, family, and social cares.

Another school teacher whose health had been impaired by years of hard work to such an extent that she had to give up teaching, began to interest herself in poultry. She made her home with a brother whose business was in one of the large cities, and residence on a farm near a suburban town. Beginning with a few dozen fowls, she increased in a few years to several hundred, the receipts each year netting her a substantial amount, until a change of residence for her brother's family was necessary to give the children the educational advantages desired, since when she has had to content herself with a few pens of fowls on a city lot.

I know a young man whom I first met as an exhibitor at numerous New England shows some seven or eight years ago. I don't know whether this young man would make a financial success of poultry keeping—make it pay—either for himself or anyone else. It has happened that he has more than once been selected by men with ample funds to take charge of poultry ventures of the kind that have never yet paid, and few men who have made poultry pay can command a better salary or show net earnings in a year larger than his salary. The situation in regard to such places as he has filled is peculiar. There will probably always be a demand for men to run poultry farms on big plans for men who will be convinced that their plans are not practicable only when it is made clear that they have not accomplished what they expected to, and they see no prospect of doing better. In my mind this problem of the man of ability, well paid for his efforts to make a success, in a sense parallels many cases we find in manufacturing and commercial enterprises — men who never really succeed in what they undertake, but who are always ready for new effort, and always find fresh opportunities opening up to them.

These opportunities must be reckoned among the possible results of one's interest in poultry as it develops.

There are also other ways in which opportunities may come to successful, or even to capable but not brilliantly successful poultrymen.

Some years ago a marketman, buying and selling poultry, became interested in the production of poultry. He began growing poultry and ducks on a very small farm, not much larger than a good sized village lot. His successes here led him to purchase a farm, and begin to build up a business on a large scale. He had his ups and downs, his business grew, but he had a heavy load to carry. Sometimes he felt like giving up — again, things looked brighter and encouraged him to keep on. The farm paid, but did not pay enough — he was not making as much at poultry keeping as he could make at something else. Then his services were wanted by a manufacturing concern for a position calling for a knowledge of the poultry business and a wide acquaintance with poultrymen. In this new line he rose rapidly. His ability and character attracted attention and brought him an offer of a position of great responsibility with commensurate emoluments.

A young man just out of college was ordered by his physician to get into some outdoor occupation for a few years. He had been interested in poultry as a boy, and concluded to go into the poultry business long enough to make a fortune and retire. In due course of time he discovered that the possibilities of wealth in poultry culture were much overrated, and that educational qualifications representing expenditures which would make a neat little capital were not at all essential in the routine work of the poultry yard. He decided that he would make more by taking up a line of work more in keeping with his educational training, and began to arrange his affairs accordingly. Circumstances diverted him from the original plan and gave him a connection with a poultry paper, starting him in a line of work which he found very congenial and reasonably remunerative.

A farmer's boy interested himself in poultry, earned a considerable part of the money to defray his expenses at college by keeping poultry, and, having graduated from college, continued poultry keeping as a means of earning the wherewithal to pay for a professional education. Shortly after he had returned to the farm he was offered the position of manager of a large poultry plant. After a few years spent here he got into poultry journalism, and is now editor of one of the papers ranked among the best. Even before his college days his contributions on poultry topics were read with interest and profit by many who might have skipped them had they known how few years the writer numbered. As another juvenile expert once remarked to a veteran whom he had bested in a discussion who tried to overawe him with years and "experience," "Some people get a great deal more experience in a few years than others do in a lifetime."

Two physicians in the same city became interested in fancy poultry. Each had a good practice. Both took up poultry for recreation. One has continued to keep a few fine fowls, getting much satisfaction and a little profit from them, but never allowing them to interfere in any way with his professional duties. I doubt whether as much as one per cent of his patients know that he has any interest in poultry. The other developed "hen fever" in a virulent form. He crowded his premises with fowls of different varieties, making all sorts of queer coops and houses for them. He neglected his practice. Patients who came to the office at his residence frequently had to wait while he attended to the poultry. Some complained that he came to them from the poultry yard with odors, feathers, or other evidence of his having been in that locality still about his person. This may have been an exaggeration. It is certain, though, that he gave many patients the impression of being more interested in his fowls than in their ailments. This was a fatal error. The ailing man or woman is of all persons most resentful of any lack of concern for their welfare — especially in one to whom they come for healing. His practice began to fall off. As his income dwindled, instead of taking steps to reestablish himself in his profession, he began to think of his poultry as a possible source of income. The last I knew of him he was still struggling to make poultry pay, and of his once fine practice little was left.

A master mechanic in a New England city has been identified with a certain breed of fowls for, I think, nearly forty years. He breeds a hundred or two every year, and makes two or three hundred dollars on them, without letting his poultry interests interfere with his regular business. He visits a few of the best shows every season for a day or for several days, as business cares permit. He meets a number of men interested in his fancy, competes with them in the shows, and enjoys their company. His sales of eggs and stock keeps him in correspondence with people all over the country. His poultry is a profitable and pleasant diversion.

A city newspaper man interested in poultry started a poultry farm in a small way, and, quite naturally, as he learned more and more of poultry culture began to disseminate his knowledge through such channels as were available. In one of these channels he found an opportunity for a special engagement out of which grew one of the leading poultry journals of the period. His connection with this and subsequently with other poultry journals, and work as a poultry lecturer have been combined with poultry keeping with results as to income which are apparently good enough to keep him from returning to regular newspaper work.

A MAN WHO FOUND HEALTH AND COMPETENCE IN POULTRY.

A printer, a young man whose health was far from robust, had to leave the office in the city and return to the farm. There the care of poultry was assigned him as "light work," most suitable for one in his condition. He became interested enough to leave the farm and go to work for one of the leading poultrymen. Continuing with him for several years he became proficient both as a poultryman and a fancier, and embarked in the business for himself. He soon made a reputation as a breeder. It occurred to him that an article he was making for use in his own yards would sell well to poultrymen generally. He began to advertise it, and soon his trade in it called for the erection of a factory. In a few years more he had a large and profitable business in the manufacture and sale of an article of a kind used in every poultry yard.

A mechanic, who was probably born a fancier, has but a small back yard in which to keep fowls. He can keep but a very few, and can raise chickens only with such difficulty that he prefers not to try to grow them at all. His method of handling poultry is rather unique. When he sees an opportunity to buy a small lot of nice fowls cheap, from someone who is going out of them, he buys. These fowls he keeps until a customer comes along who will pay a good price for them. Then he sells, and his yard is empty until another opportunity to buy low occurs. He has told me that he frequently has three or four different kinds of fowls in a year, though no two kinds at the same time, and that he makes much more in this way than he could by breeding or keeping one lot of fowls permanently in his narrow premises.

I could go on indefinitely, multiplying illustrations of what people have done and are doing with poultry. The most common type — the farm poultry keeper — has hardly been mentioned. The illustrations given nearly all refer to specialists of various kinds and grades — to people who go into poultry keeping. On the farms generally, poultry is kept as a matter of course, but as we consider the possible results from poultry in the succeeding section there will be occasion to say more of farm conditions and possibilities.

LESSON VI.

SECTION III.

Branches of Poultry Culture and Classes of Poultry Keepers Considered in Their Relation to Prospective Poultry Keepers' Expectations of Success.

THAT is a longer title than I am in the habit of making, but nothing shorter seemed to answer the purpose. I want to tell the reader — and particularly the reader interested in the possibilities of poultry culture for pleasure and profit, or for that combination of pleasure and profit which the great majority are seeking, what poultry culture offers him, the conditions which it imposes, and the limits of different branches and limitations of various individuals.

Poultrymen have to deal with four principal classes of domestic fowls, chickens, turkeys, ducks, and geese. The first class mentioned are of many times more commercial importance than the other three combined.

What Poultry Culture Offers the Farmer.

The great bulk of our poultry products comes from farms where they are produced under conditions which make the receipts for them practically all profit. This being the case it is not often that an exclusive poultry business can be conducted profitably in any section where the farms produce more eggs and poultry than will supply the local markets.

And in these sections where exclusive poultry farms are rarely successful, it is not generally advisable for farmers to go into market poultry culture on a scale that requires them to give any considerable part of their time to poultry, or that makes poultry culture more than a minor feature of their farm work. But between the conditions in which poultry is usually kept on a farm and the limits I have indicated, there is a difference which leaves room for considerable enlargement of farm flocks and improvement of farm methods, and a considerable increase of receipts, and consequently of profits, from poultry without its encroaching on either time or land which might more profitably be devoted to other uses. On most farms the best policy to pursue in developing the poultry interests of the farm is simply to make the most of farm advantages by farm methods. The best way to go about this is indicated in the stories of the several farmers who developed as poultrymen. It is a mistake for a farmer, or any one else whose interest in poultry is just beginning, to undertake to plan for a poultry business on an extensive scale. The wise way is to extend and increase operations year by year, little by little, as results and experience indicate the most profitable lines to follow.

Ordinary poultry keeping on the farm includes, almost universally, the maintenance of a flock of laying hens, varying in numbers from a few dozen to several hundred, and the rearing each season of as many chickens as can be conveniently taken care of. On most farms no special attention is given to the production of eggs and poultry for the seasons of scarcity and periods of high prices.

The profits from the poultry on such a farm may, in nearly every instance, be much increased by keeping better stock, by more careful selection of breeding stock, by disposing of poultry either before or after the seasons when most farmers are in the habit of disposing of their surplus, and in a general way by taking precautions to avoid the losses which often greatly reduce the stocks of fowls and flocks of chickens. Along these lines of saving and judicious handling there are good opportunities for greater profits with little additional expense — sometimes with no increase of cost whatever.

Beyond this, the farmer may go into any of the special branches of poultry culture as far as circumstances, inclination, and experience take him. He as well as another may be a fancier, breeder, and exhibitor of fine fowls. His opportunities for producing them are second to none, and as they are sold principally through mail orders, location cuts very little figure in that line of the business. He may grow broilers, or raise winter chickens for roasters, if his location favors, and other demands on his time permit. He may also combine, as only those located on farms can, the growing of the several kinds of domestic fowls, and keep in addition to his chickens, flocks of turkeys, ducks, and geese.

The farmer who becomes interested in poultry will see possibilities of profit in all these lines. If he develops in any direction slowly he avoids making serious and expensive mistakes. People often write me for instruction as to the best way to begin, to avoid mistakes, saying:—
"I have only a small capital, and I must start right. I cannot afford to make any mistakes."

It is not possible to altogether avoid mistakes, but there is one sure way of avoiding bad mistakes, and expensive mistakes, and that way is well stated in the old maxim, "hasten slowly." If one observes that rule he finds his knowledge of poultry and his capacity and skill in managing it growing as his stock increases, and though his mistakes may temporarily hinder him at times they are not likely to cause his failure.

What Poultry Culture Offers Those in Other Occupations.

It may safely be put down as a general rule that a person who has a business or occupation in which he is making a living ought not to change abruptly to a line of which he knows nothing, and in which his prospects are uncertain. Yet I suppose that at least three-fifths of those who come or write to me to ask how they can get started in poultry keeping belong to this class. They are for the most part persons who are dissatisfied with their present occupation, and chafing under its unpleasant features. Their interest in poultry commonly springs from impressions of it derived from hearsay or from occasional extravagant references to it in newspapers. Almost invariably they regard it as a business singularly free from drawbacks and holding possibilities of big incomes made by the hens, while the keeper takes it easy. They also believe that a poultry business large enough to give them a living income can be built up in a few months on a capital that would be too small to be of any use in most lines of business. This belief is so diligently fostered by the poultry press, by poultry writers in agricultural papers, and by those interested in the sale of equipments and supplies for poultrymen that it becomes a very hard matter to convince a man once imbued with it that it is wrong.

Unless one is so situated that he or some member of his family equally interested in poultry can give the flock all necessary attention, and can both increase the flock, and constantly increase the time given it as the growth of the flock calls for more and more of the keeper's time, it is quite useless to make a start with a small flock with the expectation of gradually developing. To do so is like planting a tree in a space which will be ample for only a few years. When the tree outgrows the space it is an incumbrance. And that is very apt to be the case with a poultry business started under conditions which limit its development long before it has reached the point where the proprietor is justified in making it of first importance.

Poultry culture has no encouraging offers to make the man already established in something else, doing reasonably well in it, and not under necessity of making some change. In almost every case where people quit other employment to engage in and learn poultry keeping they soon find that they have made a mistake. It might not be so, if they were willing to learn the poultry business before going into it, but nearly all of them are unwilling to take a thorough training. Almost as often as I suggest this course to an inquirer or correspondent the reply is:—"I don't want to do that. I learned my present business that way years ago. I am too old to do it that way. I can't afford to take the time. Can't I just begin and learn poultry keeping by keeping poultry? Doesn't my training in the other line count for anything in this?"

The reply is:—One can learn poultry keeping by engaging in it on his own account, and paying for all his experience as he goes, but it is the most expensive way to learn, and also the most discouraging, and where one persists and makes a success scores give up either from discouragement or lack of funds. Even with plenty of capital it usually takes three or four years to get a poultry plant started this way in running order and netting anything above expenses. A man who knows the business may begin with only a few hundred dollars, a good reputation, and good credit, and by hard work and the most rigid economy, both on the plant and in the household, may get along and ahead, but it is not an easy road to affluence.

What Poultry Keeping Offers Those For Whom a Change of Occupation is Desirable.

To a great many people engaged in occupations injurious to their health, or utterly distasteful to them, or in which compensation is small, and chances of advancement smaller, poultry culture seems to offer a way of bettering themselves. Yet I have found comparatively few, even of these, who were willing to work themselves into the business in ways that gave reasonable assurance of their being able to continue in it permanently. The bane of the poultry business seems to be the desire of everyone starting in it to "boss," to be his own boss if possible, if not to secure a position as manager and "boss" for someone else.

There are many poultrymen in the business, making livings of all grades, from a bare subsistence to comfortable independence and a bank balance that shows a gratifying increase nearly every year, who got into it from one of the classes we are now considering; but so far as I have known them they have worked into poultry culture gradually. Either they went to work for poultrymen and learned the business thoroughly, or they found it possible to build up their stock of poultry in the days and hours which had they had only their other occupation would have been idle. In a great many instances, too, the help of one or more of the other members of the family has been an important factor. Perhaps the most common case is where a wife equally interested with her husband looks after the poultry when he is unable to do so. Occasionally — perhaps I should say often — the wife is the better poultry keeper, and better manager of the two, though rather averse to being known to the world as the head of the concern. In most instances where women successfully carry on a poultry business in their own name the male members of the family give such assistance as they can, though occasionally a woman is left entirely dependent upon her own efforts and such help as she may hire.

In general it may be said that to make a satisfactory exchange from some other line into poultry keeping requires exceptional conditions. Many who would like to do so cannot, and some who would be better for the change could it be effected find it impossible. It should be said also with reference to poultry culture as a business for those not in robust health that their adaptability to it will depend more on whether they can stand steady work and exposure than upon ability to do hard work. A poultry keeper must be about in all weathers. He may arrange his plant to make his labor as light as possible, and to have everything adapted to the comfort of fowls and chicks; but still occasions will arise when in heat, or cold, wind, rain, or snow he must get about and look after his stock, and if he cannot stand such exposure, and cannot occasionally work long hours he had better not go deeply into poultry keeping. In some diseases long hours and exposure are not injurious, provided there is not too much overtime, and one takes proper care of himself after exposure, and in such cases poultry keeping may be unexcelled as a health restorer. The great thing is to keep the business within the limits which will not make it too great a burden.

What Poultry Keeping Offers Those With Money to Invest or Land to Occupy.

Under this heading we consider poultry culture as an investment for men who do not intend to give their poultry plants their personal supervision, or who plan to hire an expert to run the business until under his instructions they become qualified to manage it for themselves.

A great many people every year contemplate poultry culture seriously from this point of view. Men with money to invest, having heard reports of large profits from poultry, imagine that by investing money in a poultry farm they can have their money pay them ten or twelve per cent, or even more. Men with unproductive land which they are not able or qualified to handle profitably often think that by forming some sort of partnership arrangement with a poultry farmer they can get an income from the land.

All these schemes presuppose a profit on poultry large enough to yield the owner of the land, or the plant, better returns than he would be likely to get on any other investment of his money, or use of his land. As far as I am able to learn, no arrangement of this kind has ever been satisfactory — no investment in a poultry plant managed by another for a man who knew nothing of the business, has ever been profitable, and no poultry plant established on a large scale, as an investment, has ever been run at a profit. There may be instances which constitute exceptions to these statements, but I do not know of them. I do know of large plants started on a large scale, which those interested in them claim are paying, but I know of no such farm in which the statements that have been given out as proof establish the claims made. I do not know of a single large poultry plant in profitable operation except such as have been built up gradually from small beginnings by men who were trained poultrymen, most of them acquiring their training on their own plants, and conducting them for some years at first on a very small scale.

The great obstacle to the development of plants of this character is the difficulty of securing competent managers. As indicated in one of the sketches of successful poultrymen in the preceding section of this chapter, men who have the executive and commercial ability required for the successful management of large poultry plants can usually get larger returns at something else. Men who are equal only to the management of a "one man plant" almost always prefer to work for themselves, or to take charge of plants on which they are not expected to do the impossible.

It may be that the time will come when poultry plants will be satisfactory as an investment, and when men with unoccupied land can hire it advantageously to poultrymen. It may be that in the many failures of attempts to do these things we are gradually working out methods, training men, and approaching conditions which will make these things possible, but as I see the facts today I could not offer any great encouragement to those who looked to poultry culture as an investment for their funds, or a use of their land, unless they were competent themselves to manage the venture.

What Poultry Keeping Offers in Salaried Positions and Wages.

The best paid salaried poultrymen are fanciers of much more than common ability as breeders and exhibitors. They are few in number, and, as I can place them now in my mind, are all employed by well to do fanciers who cannot personally manage their plants and attend to their exhibits. By that I mean *cannot for any reason*, including more important demands as well as inability. Not a few men of means in the fancy are men whose judgment and skill, while not perhaps equal to that of those who devote all their attention to poultry, are still of first rate quality. In fact it seems to be impossible for the first rate men in salaried positions to be satisfied in the employ of men who are not keen fanciers. In these positions men command salaries from about $1,000 a year to $1,500 year, with possibly one or two going above the latter figure. In most cases men in these positions can earn something extra as judges or in other ways in line with their opportunities.

There is a considerable demand for managers of market poultry plants, generally at a much lower range of salaries. The equivalent of $100 per month is about as high as these salaries go. In a few instances higher salaries have been paid, but such engagements have rarely continued beyond the limit of the original contract. The average amount paid a man able to handle a poultry plant of one or two man capacity is about $75 per month. Good men in subordinate positions get $40 to $50 per month.

The poultryman's opportunities, however, are not limited to employment on poultry farms. There is an ever increasing demand for men having a practical knowledge of poultry culture, for places in all lines of business dealing with poultrymen, and a good many men who have not been able to make money with poultry in the ways they had originally planned, are finding their acquirements useful in these collateral lines. Not only so, but to many who have made a good deal of a success in poultry culture, these lines offer more profitable openings.

Hence, while the wages paid the poultryman at the start are not especially attractive, there are in poultry culture such opportunities for advancement to responsible and lucrative positions as open in every line to those who are competent, reliable, and industrious.

What Poultry Keeping Offers in Recreation.

American poultry fanciers and breeders, as a class, try to combine pleasure and profit in poultry culture, but we will consider here the pleasurable aspects of poultry keeping only, passing the connection with profit with the remark that those who went into the poultry fancy for pleasure have again and again found it at some time their chief or sole reliance as a means of livelihood, and been able to make a very good living at it.

The simplest pleasure poultry keeping affords is found in the production of eggs and poultry for the family table. The desire to have poultry grown on the premises, and known to be well fed and fresh when used for the table, and to have strictly fresh eggs as wanted, seems to be responsible for most beginnings in poultry keeping in town and cities. The pleasure this affords usually bears a direct relation to the results obtained. Very often it leads directly to the higher forms of pleasurable poultry culture. Of these we may consider here the few principal ones.

There is first of all the pleasure derived from the possession of fine fowls. From this there comes naturally the desire to produce fine fowls, and a keener pleasure in the sense of skill which comes with successful accomplishment in this direction. Many are satisfied to stop here. To others possession and skill suggest competition, and their keenest pleasure is derived from successes in the exhibition room, which give them reputation varying from local to world wide according to the sphere in which they compete and the frequency of their successes.

To achieve these successes someone has to exercise rare artistic and creative skill and judgment, and I think it may be said that our poultry fanciers today quite outclass breeders in all other lines of live stock in knowledge of the principles of breeding and in skill in their application. The man who goes into the fancy today finds himself pitted against combinations of art, skill, and judgment which tax his faculties to their utmost. He finds himself also brought into contact with men in all relations in life who meet on the common ground of their interest in poultry, and he finds that the peculiar combination of qualities which make the fancier are not the exclusive possession of any class of men or character of intellect.

I think that fanciers generally will agree that the frankly democratic equality of the poultry show is refreshing. Even those who in their social and business relations are disposed to be exclusive rarely display any of that spirit in the show room. On the contrary most of them seem to be able to meet every other fancier on terms of equality, and it is a rare thing to see any traces of either snobbishness or obsequiousness in the intercourse of poultrymen.

In Conclusion.

In conclusion, let me remark what no doubt more than one reader has reflected for himself, that poultry keeping offers a very wide range of possibilities. And let me add, and impress it upon the reader, that while the possibilities of what he may get are in the business, the probabilities of what he will get are in himself, and to some extent in his circumstances. The making of a competent and skillful poultryman is a slow process. When a very young man fills those specifications it will almost invariably be found that his poultry culture began in childhood and represents years of interest and application. When a man's interest in poultry does not antedate his mature years his knowledge of poultry is almost always deficient in some essentials — often strikingly so, yet by persistent application he may work out success in spite of his limitations. For men somewhat advanced in years the outcome of a venture into poultry keeping is so uncertain that I have never been able to see how I could conscientiously advise such to go into it. The prospects of their making a success of it are, on the whole, too remote to warrant the investment and the effort. Men who have passed the age when they can obtain new positions in the lines in which they have been engaged are constantly investigating poultry culture, and looking for encouragement to take it up. For men who have had an interest in poultry which by gradual extension may develop into an understanding of it having commercial value it may be worth while to try what they can do, but I am not able to recall a single instance of a man advanced in years taking up poultry culture and making a success of it.

LESSON VII.

Locating and Laying Out Poultry Plants.

THE owner of a piece of land upon which he wishes to engage in poultry keeping has to adapt the business to his location, markets, and his own ability. He makes his business fit his conditions as he understands them. If his ability and experience are small he may make many mistakes, and he is likely to make some mistakes, no matter how able or expert.

The mistakes people make in locating and laying out their plants vary both in kind and in degree. There are mistakes that make success impossible. Of this kind is the common error of buying land that is of a nature unsuitable for poultry keeping, or so located that extra expenses, due to location, eat into the profits to such an extent that the net returns are reduced below the lowest figure at which it is possible to maintain operations. In the earlier days of interest in poultry culture as a means of livelihood, the idea was land unsuitable for any other purpose was just what was wanted for fowls. Those who bought on that principle have had cause to regret it. From as far back as our poultry literature goes it has also been customary for most authorities on poultry culture to advocate sandy and porous soils that were drained well; and a site with a southerly or southeasterly exposure was considered preferable. Nowadays poultrymen are not quite as particular about those points, though they do not wish to get too far away from them — particularly when arranging winter quarters for laying stock.

As is so often the case, when these points of location were insisted upon as cardinal points there was a general tendency to try to secure them, even at the expense of other desirable features. As a result of this, a great many poultrymen have located in places where they had to contend with a multitude of other adverse conditions as well as with the faults of locations which would always be dry.

For best results most easily secured a medium soil is to be preferred. Perhaps the ideal location is one which gives a high and well drained site for the poultry buildings, but with adjacent low land that remains moist through the dryest and hottest summers, to which the runs may be extended. A sandy soil that cannot be kept in sod becomes intensely hot on hot summer days, and fowls and chicks confined to such a location cannot thrive. This has been one reason for poor summer laying and for difficulty in growing late chickens on many poultry plants.

Another point given more consideration of late years is the adaptability of the land for cultivation. Good grass or tillage land is usually good land for poultry. Not infrequently poultry will pay better on it than any other crop that could be grown. There is the further advantage in the use of such land for poultry that the poultry running on the land enrich it, and when the land by use for poultry becomes contaminated the poultry can be shifted to another part of the farm, and this will grow extra fine crops while being renovated. As far as I have discussed the matter with them—and I have talked of it with a great many poultrymen—I have found no man located on a farm with little tillage land who would choose such a place

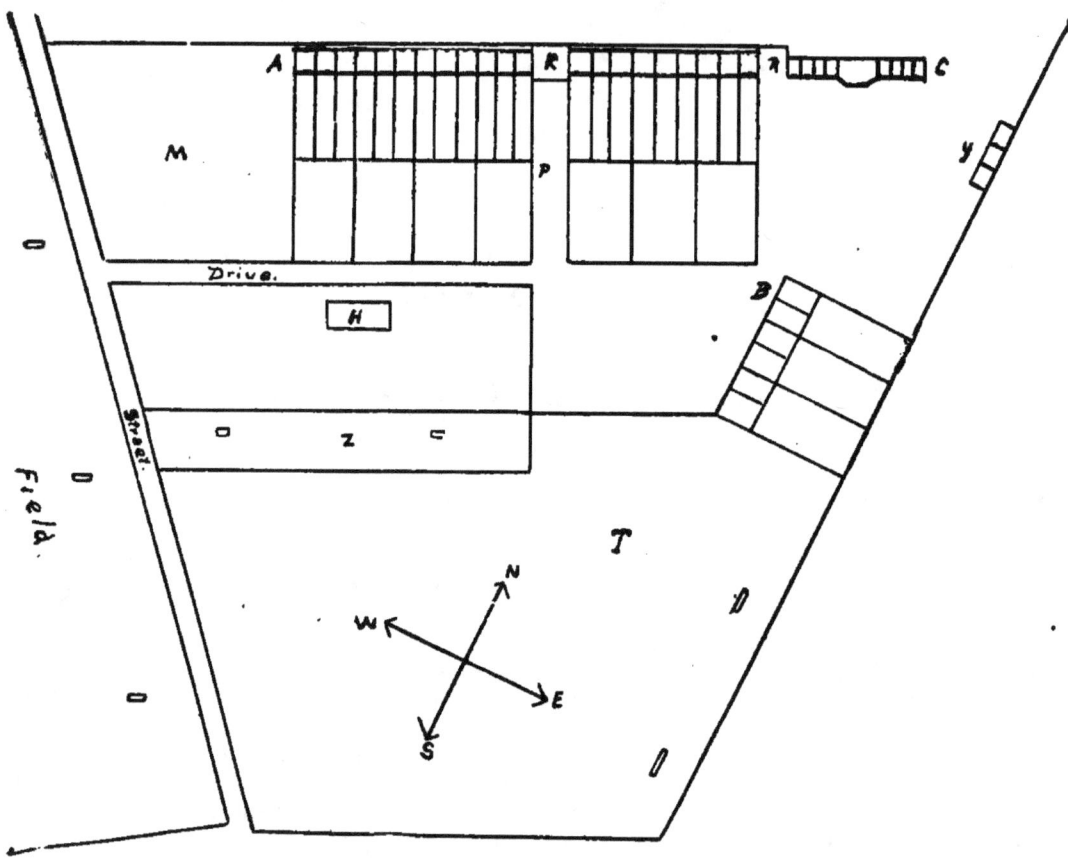

Plot of Original Grove Hill Poultry Yards.
Scale, 1-80 inch to the foot; 1-16 inch equals 5 feet.

A, continuous house with walk; K, cook and feed room, office above; R, open shed for storage; C, surplus stock house; P, approach to house A; B, cockerel and surplus stock house; M and T, large yards; Z, small yard for chicks; Y, bins for manure and refuse; small oblong figures in T, Z, and in the field to left of street, indicate roosting coops.

again, and none keeping poultry on good farming and garden land who would take land for a poultry farm which could not also be developed for grass, grain, and vegetable crops. I do not advise a beginner in poultry culture who knows little or nothing of farming or gardening to purchase land in a high state of cultivation. I do urge him to purchase land capable of being brought to a high state of cultivation.

The selection of a place to establish a poultry farm is usually a tedious process. There are so many points to consider that not a few soon begin to despair of ever getting what will suit them, buy places they do not really want, and begin to adapt their business to the farm. Only a very small proportion of the properties offered as poultry farms or farms suitable for poultry farming are desirable. Those offered cheap almost invariably have some very pronounced "outs" about them. They are hard to get to, or in undesirable neighborhoods. In general they fall very far short of the advertised descriptions in every way. But there are here and there properties most satisfactory for poultry farming which may be bought at reasonable to low prices. The finding of one of them may take months, or it may take several years. As a rule it is best not to buy until one finds a place that he is sure will be quite satisfactory. This may mean a postponement of the undertaking, but as most of those going into poultry keeping expect to stay in it, it is better to put off the beginning until a satisfactory place is found than to equip a plant on a farm with which the owner is never satisfied.

SOME MODEL PLANTS.

A farm to be well adapted to market poultry keeping should be not more than a few miles from a station that has a good express service to a good poultry market. If there is also a good local market, so much the better; but don't rely too much on a local market that has no convenient outlet to a large market, for such a market may at any time become glutted with poultry products and continue so simply by a slight permanent increase in local production.

A farm for poultry should also be of such proportions that the fowls may range widely without encroaching on the premises of neighbors, or, at least without trespassing where their presence would be objectionable. It is desirable that it should have abundance of shade, perferably orchard trees which may be expected to add something to the income, but groves and thickets serve the purpose as far as the fowls are concerned provided there is not too much wood or brush in the vicinity affording harbor for wild animals which prey on poultry. The advantage of shade already on the farm is that it provides at once good summer conditions for both fowls and chickens, while on land bare of trees and bushes makeshift shades must be provided until trees or other natural shade can be grown, and the shade supplied in that way rarely makes conditions at all approaching the best. What is usually found on such farms is a

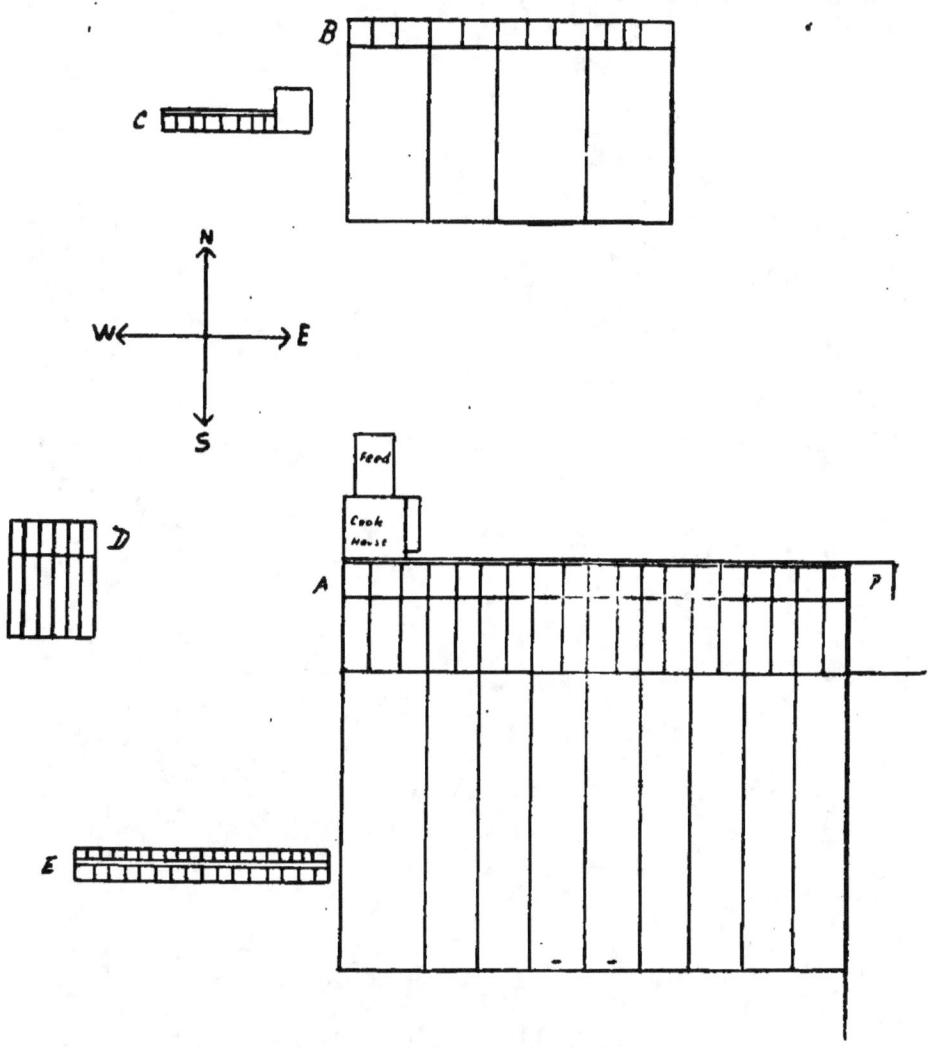

Grove Hill Poultry Yards Farm Plant.
Scale, 1-80 inch to the foot; 1-16 inch equals 5 feet.
A, 200 ft. stock house; P, shed at east end of same; B, scratching shed house; C, D, and E, cockerel houses.

A Model Fancier's Plant. Grove Hill Poultry Yards, Waltham, Mass.

GROVE HILL POULTRY YARDS.

little patch of shade here and there in which the fowls crowd for shade instead of dispersing as they should.

The land for a poultry farm should be bought, not rented or leased. In making this assertion I am aware that there may be cases where renting or leasing is better. At the same time I do not think I have ever known of an instance where it proved best not to buy — that is not after what was intended or desired to be a permanent business was undertaken.

Some Model Plants — Grove Hill Poultry Yards.

Before making any statement of methods, rules or systems for laying out poultry plants I will present plots and descriptions of several plants which will serve as models.

The first is the plot of Grove Hill Poultry Yards at Waltham, Mass., which so far as buildings and equipment go is two complete plants.

The original plant is at the foot of the hill from which it takes its name, and occupies a part of the estate upon which are the residences of Mr. Bright and of his mother. There are some seven acres in this estate, half, or perhaps a little more of it, being occupied wholly or in part by the poultry.

The estate fronts upon Main street, which is the thoroughfare between Watertown and Waltham, Grove street beginning at Main street and starting at right angles to it, but later, taking a winding course, intersects the land, dividing it into two nearly equal parts. It is on the easterly of these divisions that the two residences and all the poultry buildings, also a small dwelling for a man, are located.

The residence of the elder Mrs. Bright is at the corner of Main and Grove streets, and at the foot of Grove Hill; that of Mr. W. E. Bright is high up on Grove Hill, which at the back makes quite an abrupt descent. Half way down this descent are the dog kennels. At the foot of the hill back there is a triangular space of gently sloping land. Here it is that the poultry yards are located.

The principal building, A in the diagram, is close to the foot of the hill; in fact, in places the hill has been dug out a little for it. It is 185 ft. long, 12 ft. wide, except the central part, and contains a central two story building (K) 14 x 15 ft., used downstairs as a cook and feed room, and upstairs as an office; and two wings, one 98 ft. long, containing 12 pens, and one 72 feet long, containing 9 pens. This building faces southeast. It is lathed and plastered, and has a cement walk in the rear of the pens in each wing. A half tone from a photograph of this building appears on page 78.

On a line with the front of this building, at the end of the east wing, is a small house (C) which was probably once *the* poultry house, but it is now used for surplus stock, sitting hens, chicks, or whatever is convenient; this house is 55 ft. long, with a central pen having a hexagonal front, and two 20 ft. wings. The wings are 8 ft. 6 in. wide, and each contains four pens. The hexagonal front pen is, at the widest part, 3 ft. wider than the wings. Between A and C is an open shed, R.

Each pen in A has an outside run of th same width as the inside pen, 8 ft., and 35 ft. in length. Then for every three of these pens there is a grassy yard, 24 x 40 ft., set with fruit trees, to which the pens have access in turn. There are no separate outside runs for the pens in C, these opening in common into the yard between C, the east line of the yards of A, the east boundary fence, and the north end and fence of the cockerel house B.

This cockerel house B is a story and a half house 60 ft. long by 10 ft. wide. It runs almost exactly north and south, and has full windows in each lower pen on both east and west sides. The first floor is divided into five sections, the north one, into which the door opens, being used as a store room. Upstairs are the cockerel pens. At the east side of the house are yards corresponding to the inside pens. The fowls in the south pen also often have the run of the large triangular yard, T. In this yard T are several roosting coops for chicks. Generally a number of broods of Leghorn chicks are started in small coops in this yard, transferred later to the roosting coops, and finally the cockerels put in the end pen of C, the pullets going to other quarters.

There is a small yard south (properly southeast) of the dwelling house which is used for chicks. Then at the west end of A, between this house and its yard and Grove street,

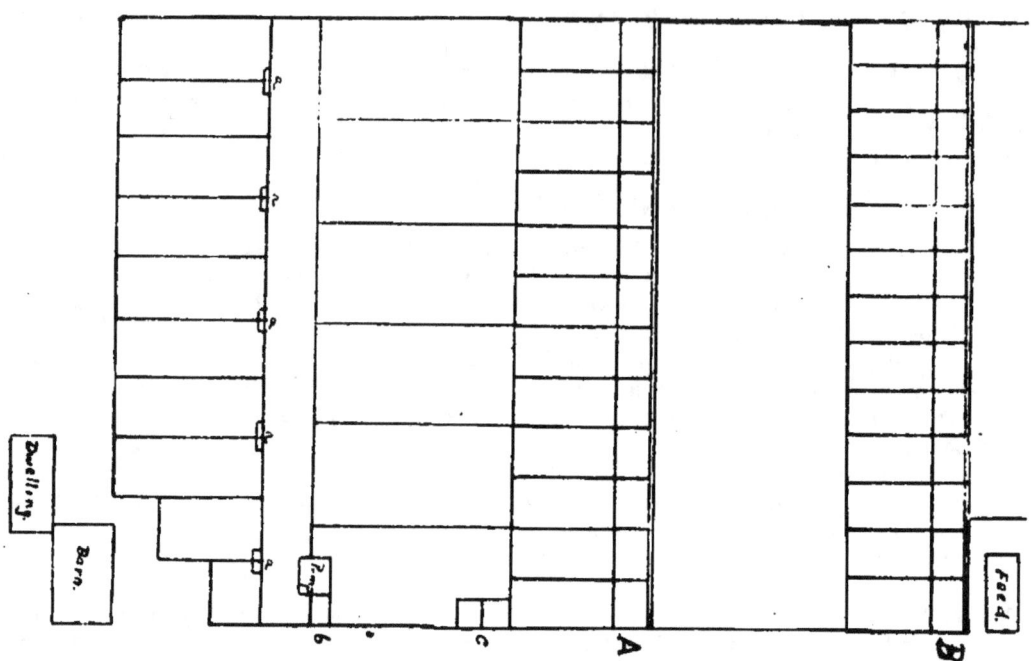

C. F. Thompson & Co.'s Poultry Plant
Scale, 1-80 inch to the foot;

A, B, C, stock houses, long houses with walks, brooder house in east end of C; is a large yard, used, as a rule, for a breeding pen. On the other side of Grove street is a pasture used for a few broods of chicks in the early part of the season, and later for pullets, these being housed through the summer in slatted front roosting coops placed in a row under the trees near the street with intervals of about 100 ft. between the coops.

Mr. Bright's Farm Plant.

On the home plant Mr. Bright had to make the buildings and yards fit the space available, but on the thirty-five acre farm, less than half a mile away, he had ample room for whatever sort of building equipment he might want. The beginning of the poultry plant on this farm was the 200 ft. house A, with cook and feed house attached.

This house faces squarely south. It is 15 ft. wide, and contains 19 pens 10½ ft. wide by 11 ft. deep. The passage in the rear of the pens is 4 ft. wide. The cook and feed house is just back of the west end of this house, and consists of one room 20 ft. square, in which are the cooker, bone cutter, pump, etc., an L, 12 x 20 ft., containing the feed bins, and a lean-to 8 ft. wide, in which is the boiler. At the east end of the 200 ft. house is a shed for manure.

The arrangement of yards here is similar to that on the plant first described, except that the large yards are longer, and there are more of them. The yards next the house, corresponding to the pens inside are 10½ ft. wide by 30 ft. long. The general plan is a large yard for every two of these, just the width of two yards, and 120 ft. long. The last long yard is irregular. As the number of small yards is uneven, it is made the width of three small yards. There are grape vines in the small yards running up over the division fences and affording fine shade. The large yards are set with fruit trees.

About 200 ft. back of the house A is a scratching shed house B, 126 ft. in length. This house was made from part of an old barn and some other out buildings, and is of such irregular construction that I did not attempt an accurate diagram — not thinking that a matter of special importance in this connection. There are about twice as many pens as yards, some having scratching sheds, and some small pens having none. These small pens are used mostly as accessory to the large pens, or for sitters or extra males. The yards in front are 72 ft. deep, and of varying widths, the narrowest being 26 ft.; the widest 35 ft.

A TWO THOUSAND HEN PLANT. 81

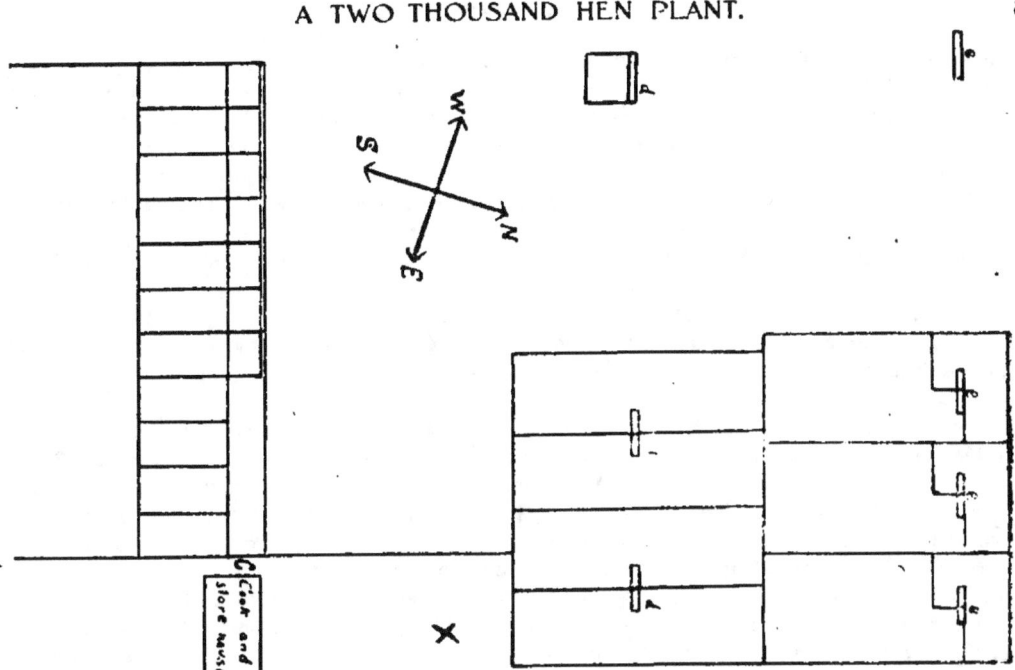

at Lynnfield Center, Mass.
1-16 inch equals 5 feet.

a, d, e, detached coops for hens and chicks, or for surplus stock; b, c, and i, small houses.

West of the house, with a roadway 14 ft. wide between them, is the building C, the main part of which is a cockerel house 11 ft. wide by 40 ft. long. At the east end of this is a shed 16 x 20 ft. for hitching place for teams. This cockerel house contains eight pens 5 x 7 ft., with walk a little over 3 ft. wide in the rear. The outside measurement of the width of the building is 11 ft. The other measurements were made inside, hence the discrepancy of some inches. There are no outside pens connecting with this house.

Directly west of the feed room and 100 ft. distant from it is a second cockerel house D, 12 x 36 ft., containing 6 pens 6 x 12 ft. Outside are yards 36 ft. long and of the same width as the inside pens.

In front of this house, and 98 ft. from the line of the front of the 200 ft. house is a third cockerel house built last fall. This house is 15 ft. wide, not quite 100 ft. long, and contains over fifty pens.

The land actually occupied by the poultry plant described, including spaces between the separate buildings and yards comprises a little over three acres. Much of the remainder of the farm is given to the young stock, the growing stock in roosting coops being well spread over it. The mowing land gives a heavy crop of grass before it is needed for the chicks. A couple of acres are planted to cabbage for the fowls every year. Some grain is grown for hay and litter, and there is some ground in garden crops, but the growing chicks have all the range they can use.

Then several hundred yards in front of the house A there is a grassy shrubby piece of low ground where several sheds are erected. In these after the breeding season the hens from the breeding pens take their vacation.

A Two Thousand Hen Plant.

The plant of C. F. Thompson & Co., at Lynnfield Center, Mass., is another case where the land, some dozen acres, allowed a liberal margin around the houses and yards, and so required no close figuring on space.

Still it is quite on the extensive plan, and while I have called it a 2,000 hen plant, and the winter capacity is over 2,000 hens, Messrs. Thompson & Co. do not attempt to grow even half their young stock here, but have over half of it grown for them elsewhere.

Viewing the diagram of this plant first as a whole, we see first a row of small houses, a, a. Then the small house b, with the pump house next it. Then back of this another small house, c. Then the three long houses, A, B, and C, and the cook and feed houses. Back of these again are more small detached houses, and far back of these a single house, i. Back of the cook house is a house which could be used for a man if required, but is now occupied by a tenant. The distance from the road in front of the residence to the rear line of the farm is nearly a quarter of a mile.

The poultry houses face the southeast. A narrow road runs from the public road in front along the northeast line of the plant as platted.

The long houses, A, B, C, are of the same width, 12 ft. A and B are each 240 ft. long. C is 200 ft. long. There is a 3 ft. walk in each house. In A there are 12 pens; in B, 13. In C the first 68 ft. from the northeast end is the brooder house. The remainder of the house is divided into seven pens.

The yards for A are 12 yards, each 42 ft. long, corresponding with the interior pens, and six larger yards with the width of two small yards. In front of B the small yards are 38 ft. deep, and between them and the house A is a large undivided yard to which the fowls from B have access alternately. There is quite a dip in the ground at this point, and for convenience in work this little valley is bridged by an elevated walk extending from the end of A to B, which saves a great deal of up and down hill work in feeding and watering.

The large cook and store house is 24½ x 42 ft., the smaller feed house 15 x 30 ft., with a root cellar under it; the small houses, b and c, are old buildings that were on the farm when purchased by Mr. Thompson. For yard room the fowls in b have the long yard running crossways of the plant back of the small houses a, a, while those in c are given the run of the large yard which should, in accordance with the plan, be used for the first pens in house A.

The small houses a, a, are each 5 x 8 ft., divided in the middle, and are used especially for hens and chicks, for the young broods first, later for the growing stock. In winter they furnish good places for surplus males. Except for the first two where the yards had to be short on account of the projection of the dwelling house and barn, these yards are 58 ft. in depth.

The small houses in the rear of the plant are each 5 ft. wide by 18½ ft. long, divided in the middle, giving pens suitable for detached brooders, for roosting quarters for growing stock, or for small pens of mature fowls. As will be noticed, the house at the extreme end of each row is out of the system, and these houses are in fact extra, not regularly used. That in the first row has a small yard in front of it; the other has none. Their principal use is for vacation quarters for the breeding stock.

It will be noticed that the yards for the pens in the southwest half of each house in the row d, d, are smaller than the others. To compensate for this, the chicks from these pens are alternately given free run in the vacant space back of them.

Something like a hundred yards back of this row of houses is a house 14 x 30 ft., which has been used for different purposes, but will next year be used for breeding pens of Leghorns. Scattered about this undivided space are coops similar to those commonly used as roosting coops for growing stock. In these the breeding hens kept over are every year given their vacation through the hot months.

A Roomy Plant for a Small Space.

By a "small space" here I mean small as compared with those we have been considering. This plant was on a three acre lot in the residence portion of a town. As described it occupied about half an acre, such a

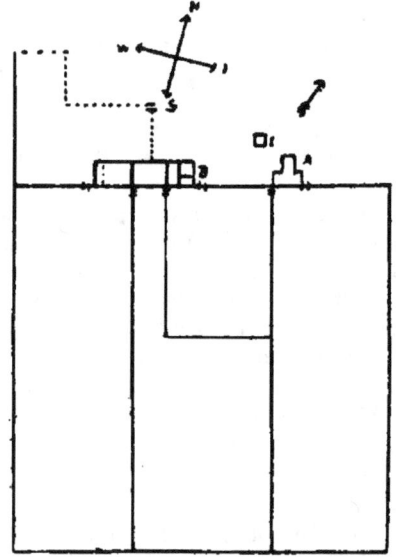

A Plant to Fit a Small Space.
Scale, 1-80 inch to the foot; 1-16 inch equals 5 feet.
A, old house; B, new house; c, bantam house.

A ROOMY PLANT FOR A SMALL SPACE.

space as it is possible to give poultry on many large village lots. I occupied these premises as lessee for several years before buying my home, and the buildings used were put up with the intention of removing them at the expiration of the lease, and were moved then, though circumstances necessitating my absence from home at the time made leaving the work to a man not familiar with the construction of the large building unavoidable and the job was badly handled.

A part of the house A was on the place when we moved there. At a cost of about five dollars a shed was added to this little house which more than doubled its capacity. This house was used in the season for one breeding pen. The yard joining it is 40 ft. wide by 150 feet deep. For shade it has a large grape vine, a small apple tree, and several pear trees.

The yard space available in front of the new house B, was not quite 100 ft. in width, and 155 ft. in depth. There were three rows of apple trees in this space, and in order to divide the shade as evenly as possible, as well as to make yard room for the fowls in the 6 x 12 pen, the principal division fence was put on the line of the middle row of trees, this being about 10 ft. nearer the west than the east wall. Then the yard for the small pen was taken in the corner of the east and larger division.

Back of A, and a little to one side is the bantam house, C, the fowls from which had the liberty of the lawn. By dotted lines in the west pen of B is indicated a partition put in temporarily at times for the accommodation of a small special mating, the yard being as indicated by the dotted lines outside.

The plots and descriptions of these plants as I have given them, indicating the positions and dimensions of houses and yards, afford but an incomplete idea of the plants themselves. In each the spaces available outside of the yards must be regarded as essential features, giving the opportunity to raise young stock under better conditions than are possible with close yarding, and also to put out the old stock at times to give it a change and rest.

I might describe a number of other plants, but cannot at present give plots of them, and in a general way the descriptions of the best of those where continuous long houses are used would be but a repetition of some of these, with variations according to the size of the plant, the "lay" of the land, and the proprietor's appreciation of the needs of his stock and the advantages and possibilities of the situation.

Where the colony plan, or any modification of it, is used, the houses are usually so much scattered that to plot the poultry houses and ranges on a scale that would give any correct appreciation of the arrangement, is not practicable for use in this connection. Besides, the colony plan more than any other, requires to be adapted to the features of the farm and to other operations carried on upon it, and the variations in it made from time to time with reference to those things may give very different arrangements in each of a series of years. In a nutshell this system consists in distributing about the farm, as is at any time most convenient, a number of small poultry houses of a capacity of several dozen hens each.

In the laying out of a permanent system of houses and yards the most important point to consider is the convenience of the keeper and economy of time and labor in caring for the fowls. In making this statement I assume that the comfort of the fowls and their needs indoors and out have had due consideration in the decisions as to the style of house to be used and the kind of yards. Of course in practice we cannot make such a separation of points to be considered, taking up one at a time and deciding it. We have rather to settle each point tentatively, then taking it up in connection with the next decide whether it can be carried out as we had planned or must be modified or entirely changed to get best results. One style of house might be preferred if each house was to contain only a few pens, while if a house of many compartments was to be built another plan might be better. The house plan preferred might require a system of yards which in some situations would be the best, because they would utilize all the land available for yards, but in other situations would not be satisfactory because they left out of use land which with another style of house and arrangement of yards could be used to the great benefit of the stock, and with some saving as well.

84 LESSONS IN POULTRY KEEPING — SECOND SERIES.

Occasionally it happens that the possibilities of a piece of ground are plainly seen at a glance, and almost anyone at all familiar with the laying out of poultry plants would know at a glance just the best way to arrange a poultry plant there. Oftener the best arrangement for the poultry plant comes to one only after a good deal of thoughtful study of the situation.

To anyone, however inexperienced, looking at a well arranged poultry plant it seems a very simple thing to lay out such a plant, but sometimes the simplest and most natural looking arrangements have been reached only after a good deal of puzzling over the situation and several remodelings. This is a feature by no means peculiar to cases such as we are now discussing.

Indeed partiality to a certain style of house or arrangement of yards has been a stumbling block in the laying out of many a plant, when if, as it was found that the preferred style of house and yard did not suit the case, the poultryman had set about making yards to fit the land and his convenience and houses to suit, the solution of the problem would not have been long delayed.

A long continuous house requires for its location a piece of ground that is level one way, otherwise the house must be built with one end higher than the other, or with short sections on different levels, either of which arrangements is unsatisfactory, the former making the temperature in the house very uneven, and the latter making passage through the house very inconvenient.

Again it very often happens that a piece of land most suitable for a long poultry house is so situated that by locating one or more long houses on it the yards are very much restricted. I have seen more than a few plants where all the advantages of giving the fowls good range had been sacrificed in this way.

Often the desire to have the poultry plant convenient to the dwelling and to other outbuildings is responsible for placing it where it is restricted on every side. Convenience in this respect is a point well worth considering, but it ought not to be secured at a loss of other advantages. It is certainly very much better to have a little longer walk between the dwelling and the poultry buildings than to expend in other ways more time and labor than is saved by having the poultry plant convenient to the house. If the poultry layout can be close to the dwelling, and convenient and suitable in every other way also, so much the better. The small plant of which a plot has been given combined in an unusual degree convenience to the dwelling and convenience in every part of the work.

For the most satisfactory layout of a poultry plant that is not made to exactly fit a prescribed space, the poultryman must take time. Even the "expert" is likely to make mistakes if he goes on a place to lay out a certain type of plant, and does so according to the situation as it appears to him at the time. Indeed, plants that are laid out by experts in this way, and, in fact, almost all poultry plants laid out on a large scale at the beginning are quite sure to prove unsatisfactory unless all features of the business can be carried on according to the original plan — which very rarely happens. The usual thing is for the business to develop along lines somewhat different from what the proprietor designed, and in this case the equipment must be changed to suit, or used at some disadvantage.

If the poultryman, as I have more than once advised in the course of these lessons, is content to let his plant grow slowly, build only as he needs, and build inexpensive buildings, he gives himself ample time to consider different plans in their adaptability to his needs and opportunities, and also to test different types of buildings and different methods on a small scale before introducing them on a large scale. This subject is one to which expert knowledge is more difficult to apply than to most of the subjects on which poultrymen ask for advice. To know a location thoroughly you must have summered and wintered with it several times over, and seen the effects of different weather conditions. So I have always been reluctant to give suggestions about the laying out of poultry plants for anything more than suggestions which would help the parties interested to solve the problem for themselves. I have laid out for myself one large poultry plant and two small ones. In no case was I able, though I was reasonably deliberate about it, and in the two later instances had a good deal of knowledge of other plants to draw upon, to make a plan that was so good I could not, after using it a little while, improve on it. This, I think, has been the common experience of those who have given

VALUE OF VISITING POULTRY PLANTS.

the subject much attention. One of the most convenient plants I ever saw finally took shape on a spot that would generally have been condemned as quite unsuited for the laying out of a model plant. The man who planned it was much above the average in intelligence and inventiveness, but the plan as ultimately worked out did not come to him all at once.

So I feel that in a lesson on this subject the most that I can do for a poultryman of some experience is to tell him what some others have done, and perhaps remind him of a few points he has overlooked. And for the beginner the most that I can do is to give him a general idea of how several good plants have been laid out, suggest for his consideration a number of points, and urge him to go slow, to take his time, not to plan too far beyond his necessities, and in his building to consider how far each part of the plant constructed can be adapted to changes. If the first buildings are of light and simple construction they may either be moved about or torn down and the materials worked over into other buildings with very little loss. If they are expensive buildings of very substantial construction they must either be used as they are or remodeled at an expense which not infrequently is as great as the cost of new cheap houses.

A most important thing for the person who expects to lay out a poultry plant is to visit as many plants as possible and study them, not so much with the idea of making a model plant, combining the best features of them all, but to find out how in the light of the owner's experience each plant answers the expectations with which it was planned. This gives one a better idea of the things that affect operations differently under different circumstances. It also gives one a better insight into the adaptability of different styles of houses and systems of poultry culture to different locations, and of the possibilities of tracts of land containing features not generally regarded as desirable for a poultry farm.

If one has the opportunity to visit a large poultry farm again and again at different seasons, he can get a very much better comprehension of the good and bad points of its plan. If he is learning the business and can make himself something of a journeyman working for a season on each of several good plants, he should be able, if he is constantly studying the subject for himself, to get something of the best out of each method and plan, and thus bring to his own plan when the time comes to make it a practical combination of good methods and features which it is rarely possible for those who plan without having had practical experience to make.

LESSON VIII.

Kinds, Breeds, and Varieties of Fowls.

BY "FOWLS" in this lesson we mean birds of the species "*gallus domesticus*," which for want of any other specific English term, are often called "chickens," though "chicken" applies properly only to their young. Perhaps in time, common usage will limit the use of the word "fowl" as we limit it here. There seems to be a tendency that way. Most poultrymen so use the word, but the explanation of this use of the term is sometimes necessary for readers not familiar with that usage.

We may classify fowls as:— Common or mongrel, cross bred, grade, pure bred, or thoroughbred, and "Standard" bred.

Mongrel fowls are fowls of no special breeding, generally a mixture containing the blood of many varieties of pure bred fowls, though in some sections there are still to be found flocks in which the blood of the old common stock is still strong. A flock of mongrels generally presents numerous very distinct types of fowls.

Cross bred fowls are the result of a union of two pure breeds. The term is usually limited to the produce of a first cross.

Grade fowls are produced by a systematic series of crosses, beginning with a thoroughbred male and females of mongrel stock, and mating each year a male of the same pure breed with females from the mating of the previous year. By this process, in the course of a few years, the stock becomes practically thoroughbred. Sometimes pure bred females are used in the first instance.

The terms *pure bred* and *thoroughbred* are synonymous, and the term *Standard bred* is also generally synonymous with the others.

As a matter of fact there are few, if any, stocks of fowls that are absolutely pure in blood; that is, entirely free from any mixture or trace of the blood of fowls not of their kind; but most of our pure or thoroughbred races are sufficiently well bred to make the production of specimens plainly showing obsolete ancestral characters extremely rare. A *Standard bred* fowl is a fowl bred to conform to the requirements of the "American Standard of Perfection," as promulgated by the American Poultry Association.

To be admitted to the "Standard," a breed or variety must be able to reproduce its type in a large proportion of its progeny. There are also other qualifications to be considered, as whether the breed presents new and distinct features; but as the judgment of the association is sometimes erratic, it happens occasionally that some fowls that are entitled to recognition are excluded, and also that unworthy varieties and breeds are admitted. Hence we find some pure breeds not "standard bred," because they are not "in the Standard;" and some breeds in the Standard that breed very indifferently. We also find breeds in the Standard in which poultrymen are little interested, while we may frequently find very great interest taken in breeds outside of the Standard.

KINDS, BREEDS, AND VARIETIES OF FOWLS.

In this lesson we limit special consideration of fowls to thoroughbred or pure bred fowls. Common or mongrel stock rarely satisfies for long the person trying to get pleasure or profit from poultry. The lack of uniformity in it, and the uncertainty of results in breeding from it, usually makes persons especially interested in poultry turn to pure bred stock to get these points, even if they have not been convinced that well bred stock are more profitable — better egg producers, and better for market poultry.

That well bred, thoroughbred stock is better for egg production, has been demonstrated in practice over and over. This statement will hold good in spite of the fact that some successful poultrymen do not use well bred stock — are not good breeders, while there is a very general belief that crosses are hardier, and many will affirm that crosses make better layers. Without entering into any argument in the premises, it may be said that there are two general facts that prove the general superiority of thoroughbreds:

First.— The great development of the poultry industry in this country followed closely the comparatively general introduction of thoroughbred fowls.

Second.— An overwhelming majority of poultry keepers whose successes attract attention keep pure bred fowls.

It would be absurd to maintain that success in poultry culture could be achieved only with thoroughbred fowls. Success depends on other things quite as much as on the kind of fowls kept. But it will be found a general rule that — whatever the cause — inability to get as good results from pure bred fowls as from mixtures is a serious handicap to a poultry keeper.

A brief reference to the history of the breeds of fowls will show how thoroughbred fowls supplanted others among progressive poultrymen, and also how some classes or types of thoroughbreds have generally replaced others. This historical showing should have a most important bearing on the attitude of the poultryman toward the various breeds and types, for in few things does the law of the survival of the fittest work more relentlessly than in the determination of the status of a variety of fowls.

Buff Plymouth Rocks.

It was between 1840 and 1850 that the American public began to be interested in the improvement of domestic fowls. Prior to that time the fowls of the country were not of a class to arouse enthusiasm, or to suggest the possibilities of development which have since unfolded. We may assume that there were scattered here and there throughout the country flocks of well bred fowls — some developed crudely by the fanciers born before their time, and some, perhaps, brought in by people coming from foreign countries where superior types of fowls were to be found; but the public generally was not attracted by them, and they made no distinct impression on the poultry culture of the time. Early in the "forties" some gigantic fowls imported from Asia began to attract attention. Traditions which are as accurate as much that passes as history, say that various lots of these fowls were brought from the Orient by sea captains. Some who do not credit the detailed stories of the importations of particular lots of these fowls, claim that they

LESSONS IN POULTRY KEEPING — SECOND SERIES.

were developed in this country and smuggled aboard the vessels reputed to have brought them from abroad. Such an explanation, however, merely removes the importation farther back— for it is not reasonable to suppose that fowls of this class and type as introduced to the public were developed from our common stock or developed without attracting attention.

These Asiatic fowls were of the general type of our present Cochins and Brahmas, but without the finish of form and feather, and without the sharp differentiation into varieties which has since taken place. They were large birds. Their great size as compared with the common fowls

Silver Laced Wyandottes.

seems to have been their first point of attraction. They were docile and hardy, and were generally given, by those who tried them, the reputation of being good layers. They laid large brown eggs—then a novelty in our markets, and as poultry were considered very superior to the common fowls.

They were immediately taken up both by "fanciers" and by poultry keepers. The fanciers immediately began to multiply breeds by giving different names to different types and colors. Quite a number of poultrymen began at once to try to improve the common stock in their hands by crossing the large males on it. It is said that one effect of this was that within a few years the poultry brought into the Boston markets was noticeably improved in size. Another result was a general quickening of interest in better poultry. People began to try to learn something of established breeds of fowls; numerous importations were made from England especially, though there may have been some from continental countries. Within a few years most of the breeds having any vogue in England were pretty well represented here, and forty years ago Hamburgs, Polish, and Spanish were distributed quite generally throughout the northern states—not in such numbers as are found of popular fowls today, but still numerous enough to become familiar objects. Asiatics seem to have been distributed more slowly. All these worked into the common stocks of the country until, when I was a boy beginning to be much interested in poultry, a large proportion of the farm flocks contained many specimens showing unmistakable evidence of well bred parentage of some of these races.

Such breeding, however, was indiscriminate, and led to nothing definite. None of the new types produced were able to gain more than local prominence. Nor did the thoroughbred fowls of those days take with the public, especially the farmers, as did those which were to be brought out later.

It is less than thirty years since the first of our present "American class" of fowls was introduced to the public as the "Plymouth Rock," to be known later, as other varieties of the same breed type appeared, as the Barred Plymouth Rock. Into the disputes with regard to the origin of this fowl we need not enter here. Suffice it to say that in it were combined for the

POPULARITY OF THE NEW TYPE OF FOWL.

first time with promise of race permanence the most generally desirable qualities of the Asiatic and the smaller races of fowls. Soon afterward the Silver Wyandotte appeared. This was a combination of the Hamburg and Brahma, and a most attractive intermediate between those types.

A few years before the introduction of these "general purpose breeds," the Leghorns had been introduced, and they quickly distanced the older "every day layers" in that field, while with the new Plymouth Rocks and Wyandottes they awakened a general interest in poultry culture which has enormously increased the volume of our poultry products, and is still steadily growing.

The remarkable success of the new class of fowls led to an immediate multiplication of varieties of the same type — if indeed some of these were not already making when the pioneers of each breed appeared. In a few years more we had White Plymouth Rocks, Golden Wyandottes, and White Wyandottes. Then came Buff Plymouth Rocks and Buff Wyandottes, and after them Partridge Wyandottes, Silver Penciled Wyandottes, Columbian Wyandottes, and Partridge and Silver Penciled Plymouth Rocks. Of the duration of popularity of each of these, and of their relative popularity there will be occasion to speak a little further on.

S. C. Rhode Island Red Cock.

Meantime the developments of this favorite type of fowl have not been limited to the varieties of the two breeds mentioned. In a section of the state of Rhode Island which still preserves a degree of isolation rare in these days of easy communication, the introduction of Asiatic fowls half a century or more ago marked the beginning of the making of a "local" breed, the only one which has made an important place for itself in this country, the Rhode Island Red. These are fowls of the general type and characteristics of the Plymouth Rocks and Wyandottes. As commonly bred before fanciers became interested in them they were not pure bred fowls, for foreign blood was often introduced, one farmer using a male of one breed, another of another breed, and so on. So general was the mixture that in most flocks of R. I. Reds a few years ago evidences of a very mixed ancestry were conspicuous. But through all these mixtures a common type was followed, and when fanciers took up the breed it required only a few seasons of careful breeding to make them as "thoroughbred" as most breeds have been within the same time after their introduction.

In England the success in America of the medium sized general purpose type of fowl led to the development of a breed of fowls much the same in type, a little more "beefy," as English types of poultry usually are when compared with American, but still very like. This breed was called the Orpington. The object of the originator, as repeatedly stated by himself, was to make a breed of the general type of the Plymouth Rock and Wyandotte, but better suited to English tastes and markets. The American productions had the yellow legs and skin popular in American markets—the English markets wanted a white skinned fowl with flesh colored legs.

In considering the relative merits of the many breeds and varieties in this class of general purpose fowls, the reader should always bear in mind that they are essentially very like; that the differences between them are mostly superficial; that in many cases differences observed between the lots of two different breeds or varieties of this class when compared are peculiar to the case under consideration, and not general differences running all through the variety or breed. Hence as we shall see, any one of these varieties may be substituted for another in any

90 LESSONS IN POULTRY KEEPING—SECOND SERIES.

S. C. Brown Leghorns.

case where superficial differences are immaterial. By a superficial difference I mean a differing not affecting productiveness or development, or the actual adaptation of the fowl for the purpose intended. Such a superficial difference may be of great importance to one poultryman, and of no importance at all to another. For instance, the color of skin — except in so far as it is an index of good condition, and that lies in quality of color rather than in any particular color — has nothing to do with quality of flesh — yet in this country people generally prefer yellow skinned fowls, and will buy them more readily, and often will pay a premium for them. Therefore if one is growing fowls to sell for table purposes, this superficial matter becomes for him an essential matter, meaning easier sales and better profits, with no difference in cost of production. If he is growing fowls only for his own table, and has no prejudice in the matter of color of skin, it will make no difference whether his stock is yellow skinned or white skinned. Or, again, take the color of plumage: A white or buff fowl is easier to dress in the pinfeather stage than others. If one is selling much poultry this slight difference in time and cost of dressing may amount to a large item in the course of a season, so large an item indeed, that he will prefer to keep only white or buff fowls; but if one is dressing only for his own use, or dressing for sale in small amounts, this advantage of color may not be of importance enough to call for consideration.

With these two illustrations we will pass the subject of superficial differences. I think that what has been said about them will enable the reader to see the point without further explanation, in each case of this kind to which allusions will become necessary as we discuss the relative merits of the varieties of fowls.

Returning to our main subject: In the Plymouth Rocks, Wyandottes, Rhode Island Reds, and Orpingtons, (of which there are some eight or ten varieties), we have a common type of medium sized, active, hardy, vigorous fowl, a good egg producer, and a good table fowl; we have now varieties representing quite all the colors, and combinations of colors, that have ever been popular in any breed of fowls. In other words, we have now fowls of this class to suit every taste in color.

Now on their economic merits and their general adaptability, this class and type of fowl, in the comparatively short time since its introduction, has far outstripped all others combined. Not only so, but in each breed in this class, (with the exception of the Rhode Island Red, of which there are but two varieties, differing only in comb), one or two varieties have gained a long lead on all the others. This means that it is these varieties that are best adapted to the needs of the greatest number of poultry keepers.

Then in the choice of a breed the beginner who is wise will not look through the entire list, and endeavor to select for himself on the descriptions he may find. He will rather consider first the most popular varieties, assuming at the outset that it is best to take one of these unless for some special reason another kind of fowl is to be preferred.

CHARACTERISTICS OF PLYMOUTH ROCKS.

I suppose that for nine-tenths of those who keep poultry, a general purpose fowl is the best fowl — will be found most satisfactory. What proportion of this nine-tenths are best suited with the most popular varieties of this class, is not so easy to estimate, but I should say certainly three-fifths, and possibly as much as four-fifths. These estimates will indicate to any reader interested in the matter, something of his "expectation" of suiting himself better by going outside of the popular varieties of the general purpose type of fowl.

But while the advantage in general popularity is overwhelmingly with one class of fowls, and with a few varieties in this class, there is still a large field for the others, and if the tendency is for growers of poultry for economic purposes to concentrate on a few varieties, the tendency among fanciers is quite opposite, and the number of people open to interest in other varieties is always great enough to make possible a liberal demand for any fowl that has merit and beauty.

Coming now to a detailed consideration of the varieties mentioned, we have *Plymouth Rocks*—Barred, White, and Buff, (which are "Standard," and well established varieties), and Partridge and Silver Penciled, (which are recent introductions, whose status is as yet not determined).

Indian Game Hen.

Of these, the Barred Plymouth Rock in this country has long led all varieties in popularity. Until a very few years ago it was probably correct to say that — counting both special poultry farms, farms on which some attention was given to making poultry profitable, and fanciers' plants, there were more Barred Rocks in the country at large than of all other thoroughbred fowls combined. Their great predominance in numbers was due first of all to their being first in the field. They had two superficial faults: They were, and still are, very difficult to breed to a high excellence in color; and they had many black pin feathers when dressed in the pin feather stage. But in spite of these drawbacks, they long held their preeminence, because the early White Plymouth Rocks were lacking in vigor, and the White Wyandottes, which, in some sections, surpassed the Barred Rocks in popularity, were, on the average, smaller fowls, and many breeders had injured their stock by forcing early egg production.

White Plymouth Rocks are now fully equal to the Barred in every economic quality, and have the superficial advantage of color, which is gradually bringing them to a popularity more nearly equal. As between White and Barred Rocks, choice hinges on taste, and on whether easier preparation for market is an advantage. It is when many chicks are to be dressed in the pin feather stage. When stock is not dressed until mature — as is the case on most farms — the white color is no advantage.

Buff Plymouth Rocks are quite popular, but not so much so, nor with such promise of permanent general popularity as the Barred and White. Their most serious fault, from the average breeder's point of view, is that common to all buff and red fowls — the wide departure from approved shade of color even when bred with great care and good judgment to maintain color. To the fancier this may not be a fault. Those who admire buff fowls, and find pleasure in producing them, will readily sacrifice the culls, but a poultryman not especially interested in the fancy will not long be suited with buff or red fowls if he wants uniform appearance in his flocks, for he cannot afford to sacrifice off colored hens. If he is indifferent to lack of uniform

Houdans.

appearance he will not count this fault against them.

Partridge and Silver Penciled Plymouth Rocks are very handsome fowls, but at present require great skill in breeding. Like all laced and penciled fowls, they will always be difficult for most breeders to handle, and if their popularity, as compared with the other varieties of the breed, follows the same course as that of similar colored varieties in the old breeds, they are not likely to come into a general and lasting popularity, but will be bred mostly by those who want general purpose fowls with single combs, and have a decided preference for one of these colors.

In *Wyandottes* we have, naming them in the order in which they were admitted to the "Standard," the Silver Laced, Golden Laced, White, Black, Buff, Partridge or Golden Penciled, Silver Penciled, and Columbian.

The typical Wyandotte, as compared with the typical Plymouth Rock, is a shorter bodied, compact, blocky fowl, and a little smaller fowl. But many breeders of Wyandottes breed them as large as Standard Plymouth Rocks, and larger than the general run of Plymouth Rocks, while it is probably true that Wyandottes as they run are smaller than Plymouth Rocks as they run. The difference in size is one which any breeder of either variety who chooses to do so may easily overcome breeding Wyandottes to Plymouth Rock weights, and vice versa, so that for general use we may say that there is no material difference in size between Rocks and Wyandottes. The real differences in this respect are, that the most symmetrical fowl in each breed is likely to be the fowl about standard weight and size for the breed. Wyandottes of Plymouth Rock weights are apt to be a little coarse and "cochinny;" Plymouth Rocks at Wyandotte weights a little undersized. From this it follows that if a poultryman wants fowls which we may term small medium in size, he can get them in Wyandottes without departing as far from a good type as he would if he tried to breed Rocks to the same weights. Conversely, if he wants large medium fowls he can get them in extra large Plymouth Rocks, which will be a little coarse, but not so far away from the breed type as if he tried to get as large fowls in Wyandottes. In egg production there is practically no difference in average number of eggs laid. The Wyandotte eggs average a little smaller, as would be expected, but the difference in this respect is not important.

As between the different varieties of the Wyandotte, the White is by far the most popular, ranking as one of the few most popular fowls. The Buff, though far behind the White, comes next in popularity, and so far these are the only varieties of the breed to gain and hold any great popularity, though all the laced and penciled varieties have had their booms, and have maintained a following large enough to keep them well in the public eye. The Black Wyandotte has never attracted much attention. The Columbian Wyandotte is by many regarded as a "coming" variety. Being a white fowl with black points like the Light Brahma, it will dress like a white fowl.

Of all these varieties the White furnishes the greater proportion of specimens of good size and with true Wyandotte shape. This is because more people are interested in it, and more of those interested in it are especially interested in market poultry culture.

Of *Rhode Island Reds* there are, as has been said, two varieties differing only in the shape of the comb, one having a single, the other a rose comb. Taking the average Reds as we find them, the single combed variety might — on its looks — be termed a Red Plymouth Rock, and

THE MEDITERRANEAN CLASS.

Silver Spangled Hamburgs.

the rose combed variety a Red Wyandotte. Fanciers of Reds try to make the Rhode Island Red of a shape intermediate between Plymouth Rocks and Wyandottes, but as the reader who examines many specimens of the three breeds will find, the breed shape is still very rare in the Reds. As a matter of historical fact, the first Buff Plymouth and Buff Wyandottes exhibited were Rhode Island Reds, and a good part of the stock of both these Buff varieties is of Rhode Island Red origin.

In Rhode Island Reds, then, we have two more varieties of the medium sized general purpose fowl, not essentially different from Plymouth Rocks and Wyandottes in practical qualities.

It is from the varieties of the three breeds just described that most poultrymen will make choice of the fowl that suits them, and whatever else they may begin with or try, most poultrymen will eventually settle on a variety of this class. All these breeds have, with the exception of an occasional stock, the brooding faculties active. All lay tinted eggs varying from rich creamy color to very dark "brown." All are easily handled, and give fair to good results when fed and cared for with ordinary good judgment and regularity.

The Buff Orpington is the only variety of that breed at all generally introduced into this country. Without denying it as great economic merit as any variety of the American class, and admitting that as seen on exhibition Orpingtons have generally shown better table form than the American varieties at the same shows, one is quite safe in predicting that their general effect on the varieties with which they come into competition will be to improve rather than to displace them. The other varieties of Orpingtons, Black, White, and Spangled, have attracted little attention here outside the circle of enthusiastic Orpington exhibitors. Considered as a commercially popular fowl, the question of the popularity of Orpingtons in this country turns on the question of the continuance of American prejudice in favor of yellow skinned table poultry. If one can convince himself that this is passing he may see a large popularity coming for the Orpingtons. Otherwise he is likely to believe that the American varieties will continue to give general satisfaction.

Next to the American class in popularity comes the Mediterranean class, comprising Leghorns, Minorcas, Spanish, Andalusians, and Anconas. These all lay white eggs, and are nonsitters.

Leghorns. In these there are seven varieties, of which only two have a broad popularity — the S. C. Brown and the S. C. White. Of the two the White seems to be most popular throughout the territory which contributes New York city's supply of fancy white eggs. Throughout the rest of the country the Browns are generally more numerous. The Single Comb Buff Leghorn made very rapid advances in popularity for a while, but then went backward. The rose combed varieties of the colors mentioned have never approached the single combs in popularity, though the Whites and Browns are quite well distributed. Black Leghorns are not often seen, and the Silver Duckwings are still more rare.

The Leghorns' chief claim to attention is their laying propensity. They lay better under indifferent care, except in early winter, than any fowls not of their class, and except when frost is severe enough to affect their large combs they are reasonably hardy. Average Leghorns are too small to be of much value as market poultry. Many breeders breed to a size to mate their Leghorn chicks, and young hens compare favorably with ordinary stock of the American breeds, but the average Leghorn is a very poor table fowl.

The Black Minorcas are the next breed in this class in popularity, though far behind the popular varieties of Leghorns in this respect. Ordinary Minorcas as found distributed throughout the country do not differ greatly from Leghorns. Indeed it is not an unheard of thing for breeders to supply customers with Black Leghorns and Black Minorcas from the same pens. Typical Black Minorcas, as bred in the sections where they are most popular are medium large fowls, sometimes as large as Plymouth Rocks; are good layers of very large white eggs, and are fine table fowls for home use, though their white skin and dark legs are not in favor in most markets.

Andalusians and Anconas do not differ much in anything but color, from Leghorns. The Andalusian is a trifle more on the Minorca type; the Ancona on the Leghorn type. The Andalusian is a slaty blue in color, and very difficult to breed to standard color requirements. The Ancona is a mixed (speckled) black and white fowl. Both have their admirers, and the Andalusian in particular is given a good deal of attention by fanciers. They may be rated as fowls for the fancier and amateur rather than for those looking for the most suitable fowl for commercial purposes.

The Black Spanish are practically extinct, except in the hands of a few fanciers.

Of much the same general type as the Leghorns are the Hamburgs and Polish. Varieties of both breeds were popular before the introduction of the Leghorns, but being generally less hardy, more difficult to keep and rear, and lacking the yellow legs and skin which our markets prefer, they were rapidly crowded into the background on the advent of the Leghorns. They are still bred by many fanciers, and seen in considerable numbers at some shows. Most Hamburgs are so small as to be of little practical value. A few breeders maintain good size, and a type that is well suited both for egg production and for the family table. The Polish fowls, though small, are generally plump and meaty. They are good layers under favorable conditions, but their large crests make them very susceptible to colds and roup when exposed to wet weather.

In the Asiatic class we have another group of fowls generally crowded out by the "general purpose" fowls. The Asiatics, of which there are three breeds — Brahmas, Cochins, and Langshans — are large fowls — too large for general market demands, and among those keeping fowls for commercial purposes are bred mostly for special markets. With the exception of the Light Brahma, it may be said that the Asiatics are of no economic importance today, and it occupies a very limited field. Light Brahmas in a few localities are produced in very large numbers for large roasting fowls. They are the largest of fowls, and as it takes them so long to attain full size, they remain soft meated much longer than fowls of the smaller breeds. With judicious management they are good layers, but comparatively few poultrymen succeed in getting satisfactory egg yields from any Asiatic fowls. The Langshan, the smallest of the group, is the best layer under ordinary management, but its color, (black, the White Langshan has never become well known), is against it for market poultry. An objection, in most sections, to all Asiatic fowls, is the foot feathering. Wherever the soil is heavy and there is much wet weather, or where these fowls are not provided with houses where the floors are dry and littered with material that will quickly absorb the water in the foot feathers after they have been out on wet ground, this foot feathering is really a fault. Asiatics are the most docile of fowls and the hardiest, but unless one gives them plenty of room and uses judgment in handling them he will not get as good results from them as from fowls of the American class for any purpose. In the hands of those who understand them they are good layers — comparing favorably with any other breed, but the average poultry keeper gets very few eggs from them, and soon changes to a breed easier to handle.

In our "Standard of Perfection" Dorkings, Red Caps, and Orpingtons are grouped together in the "English" class, though the three breeds are of distinctly different types. The Dorking is an English production of great antiquity. The Red Cap is perhaps best described as a fowl of the Hamburg class bred to large size, while the Orpington, as has been seen, is an English translation of the type which prevails in the American class.

Red Caps are rarely seen in this country. Dorkings are found in considerable numbers at some of the leading shows, and in Canada are quite extensively kept as farm fowls. They are commonly reputed indifferent layers and rather delicate. I kept a small flock secured from a Canadian breeder for several years, and in this limited experience with one stock found them hardy and good layers. As table poultry the Dorking has long ranked as of finest quality, and fully deserves its reputation.

Houdans are the only French breed well known in this country. They are as good layers as Leghorns, and first class in table quality, but not as rugged as is desirable for fowls for general use. Like the Polish, they have heavy crests which to many are objectionable. The color of their skin (white) and their dark legs are also against them as market fowls.

In Game fowls we have three distinct types — the Pit Game, the Exhibition Game, and the Indian Game. The Pit Game fowl is practically the Game fowl as it has been bred for centuries in England, with perhaps a little more differentiation in colors. The Exhibition Game is a long legged, long necked, exaggerated Pit Game, which has no place outside the exhibition room and the yards of the fancier. Pit Games are really valuable economic fowls as far as productiveness and quality go, though not to be classed for general purposes with the breeds of the American class. Perhaps their greatest fault from economic standpoints is their "gameness," — their pugnacity, and quarrelsomeness. These are qualities destructive to comfortable and profitable work with poultry, and the Game as a farm or practical fowl, soon disappears from sections where the economic value of a fowl becomes the first consideration with poultry keepers.

The Indian Game is a larger and meatier type of fowl than either of the others, and, as bred in England and America, is less pugnacious. I think it should be regarded as a useful type temporarily somewhat neglected because of the disappointments which followed its introduction to the American public, with widely exaggerated reports of its laying and table qualities.

Of course it is out of the question to discuss in a single lesson exhaustively the qualities and adaptabilities of all these breeds and their several varieties. As I said near the beginning of the lesson, the poultry keeper should limit consideration of varieties to the few popular general purpose breeds unless there are special reasons for not doing so. The most general illustrations of exceptions to this rule may be found where one is going into some special branch of poultry culture, as the production of white eggs for the New York city trade, or the production of large roasting chickens for the Boston market. Even in these exceptions, the principle of the rule I have given applies, and the poultryman should follow custom, and take the breed, or one of the breeds most popular among those in the line in which he is engaging.

LESSON IX.

Stocking the Poultry Plant.

IN THE last lesson the different varieties of poultry were described, and their adaptability to different conditions and purposes discussed. In this lesson we take up a number of questions in which beginners, wherever located, or whatever their objects, are about equally interested.

How Many Breeds or Varieties Should a Poultryman Keep?

Those who have been long in the business are generally agreed that one variety is better than more — is enough. Even those who keep several or many varieties are quite unanimously of the opinion that it is better to start with a single variety, and to limit oneself to that one variety. It does not necessarily follow from their taking this position with reference to what it is best for one beginning now to do, that it is also better for one who has several or numerous varieties of fowls to drop all but one.

Granted that it might have been better not to keep more than one variety, it still is true that when the thing that was not best has been done, conditions are sometimes created which make it better policy to keep on as one has begun than to change to the more approved situation.

A breeder of several varieties who has established a trade in each, cannot drop any of them without letting go trade which it cost him something to secure, and he cannot afford to let such trade go unless he is reasonably sure that increased sales from the variety he retains will compensate for the loss of trade which follows the dropping of the others. The beginner's situation is different. He can take the one variety of his choice, and concentrate all his efforts on securing a fine stock of that variety, and building up a trade in it. If he is successful in the first, and has ordinary good business ability, he can hardly fail to succeed ultimately in the second. In the earlier days of poultry culture it may sometimes have been good policy to start with several breeds; as to that, opinions differ. But of late years trade tends more and more to go to "specialists" — that is, to breeders making a specialty of a single variety, and in the stronger competition for exhibition honors, and for business that exists today, the man of several breeds is more apt to be crowded aside by competitors, and neglected by purchasers. He finds it harder to win a satisfactory share of the prizes where in each variety he keeps he has to contend with men of equal or greater skill in breeding who are applying to that one variety as much skill as he has to divide among two, three, or a half a dozen, and to add to his difficulties, buyers generally prefer to buy of the man who keeps but one breed. The reasons some have for doing this are fallacious, but it is the condition of which the breeder must take account. He can adjust his business to conditions much more readily than he can change conditions to fit his ideas of how business should be done.

Beginners often think it advisable to keep two varieties of different classes and types to meet different demands or serve different purposes. The most common cases are to keep large fowls for table purposes and small fowls for laying; and to keep fowls of the Asiatic or American breeds for winter layers, and Leghorns for summer layers. Usually they find that results do not justify the arrangement. The special adaptabilities of the different breeds to different purposes are more theoretical or fancied than real. Thus Leghorns, though easier to get eggs from than heavier breeds, and generally steadier layers in summer because they are non-sitters, frequently surprise the man who would keep them for summer layers, by laying as well in winter as his supposed winter layers, while the difference between the two kinds for the year may be insignificant. The fact is that in general the different breeds lay about alike when given good care — such as those trying to make poultry pay commonly give their fowls, and in time the poultryman realizes that the few practical advantages of keeping two types of fowls are about offset by the disadvantage of having to maintain two stocks, and the frequent inconvenience in adapting the accommodations to the different habits of the fowls.

What Quality of Stock?

The next question of interest to the beginner is the quality of stock to buy. Beginners usually purchase low priced stock — that is, low priced from the fancier's standpoint. From the beginner's point of view, two or three dollars for a female, and three to five dollars for a male is extravagance. He may pay such prices, but prefers not to let his acquaintances not much interested in poultry know the amounts. Occasionally a beginner will pay much higher prices. If financially able to do so, beginners often buy the highest priced birds. Their idea is that by so doing they buy a place and a standing among the foremost breeders.

There certainly is an advantage in buying high quality stock, and it may be from every consideration the best policy for one who is financially able to do so, and who knows how to maintain its quality. It is on the latter point that most beginners fail. No money can buy skill in breeding except money which may be paid to a breeder for birds he has produced, or as salary. His goods and his services may be bought if he is willing to part with them for a consideration, but this kind of knowledge and skill is not to be had separate from individuals who have it.

As we saw in the lessons on breeding, it requires quite as careful selection to maintain excellence in fowls as it did to secure it. In unskillful hands the best of stocks are apt to deteriorate rapidly. A single season of unskillful management of the breeding stock may put the stock of the man who bought the highest priced birds he could get on a level with that of one who bought much cheaper stock. Unless one is in a position to get expert services in mating his stock it is as well for him not to pay extremely high prices. I would not say that he should limit himself to the lower figures I have mentioned. He might go several times as high, and if he proves an apt student of mating problems and successful in growing chicks, have results that justify the larger expenditure for stock, but as a rule the poultryman who buys extra good stock at the start does not establish his stock on that foundation. The rule is that in his inexperience and unskillfulness his first stock goes back, and when he realizes this and sees where and why he failed he buys anew for foundation stock, and on this stock bought when he has a measure of experience to show him how to use it, he builds up his permanent line. Hence in buying ordinary good stock at about the range of prices mentioned, the beginner with poultry is simply applying the common principle in use wherever people work with materials which may be damaged or lost in manipulation, of using rather cheap material to experiment with.

Beginning With Stock or Eggs.

The determination of this question is settled in part by the season when the beginning is made—people are going into poultry keeping at all seasons. Unless the start is making during the late winter or spring, that is, during the hatching season, beginning with eggs would not be considered (except in the hatching of winter chickens, in which line, as stated, most growers buy their eggs for hatching). For those beginning at times when the start might be made with either stock or eggs, it is often a puzzle to decide which way to begin. Results by either method of starting are so variable that one has to be cautious about making positive recommendations, but I believe that the greater number of satisfactory beginnings are made from stock. Perhaps as good a way as any is to try both ways, — divide the amount available for the purpose, buy a few fowls, and invest the rest in eggs. Though there is no sureness about results, unless luck goes entirely against the beginner, he is likely to get some good chickens from fowls mated as they were sent him by the breeder from whom he bought them. But in buying eggs there is always the possibility of getting a good hatch and a lot of exceptionally good chicks at a cost away below what birds of the same quality could be bought for at maturity. This chance is attractive enough to make most of us risk the total failure which comes to the buyer of eggs about as often as a satisfactory hatch of chicks that turn out well.

Buying young chickens is much like buying eggs except that the uncertainty of hatching is eliminated as far as the existence of the number of chicks desired is concerned. The chicks are shipped before it begins to appear whether they would generally live and thrive, and the results at the end of the season are likely to average only a little better than with eggs. That little, however, is an inducement to many to buy new hatched chicks rather than eggs.

It is not possible to eliminate risks in starting, or — for that matter, at any stage of the proceedings. Whichever way one elects to begin, or if he prefers to try them all, there is risk of failure to succeed in any of the attempts.

Nowhere is persistence more necessary than in efforts to get a start with stock of poultry of the kind one desires. In what follows I shall try by suggestions and advice to help each beginner to avoid mistakes, but I cannot assure him of any way of certainly avoiding them. I made the common mistake myself, of beginning with a number of varieties, and made no more than the average number of mistakes in buying, yet it took me two to three seasons to get a good start — just a start with a few birds in most varieties, and some it took longer than that.

In many cases the question as to beginning with eggs or stock is better answered by considering it with the question,

Where to Buy.

The greater number of novices in poultry culture seem to think they can buy better stock, whether in birds or eggs by sending to some breeder at a distance for it. They see the faults — some of them — in the stock of nearby breeders whose yards they visit. The stock they do not see they judge by the breeder's advertisement and descriptive literature which rarely admit that the stock has any serious faults, and his correspondence, which only occasionally refers to the weak points in the stock, and then minimizes faults more than a disinterested person would. Common sense might teach even the novice in poultry transactions to discount liberally the salesman's enthusiastic recommendations of his goods, but apparently only experience in buying teaches this lesson effectively, and the average beginner in buying poultry will pass by his neighbors and cheerfully pay a little higher price, plus a heavy express charge, for stock no better than he could get close by.

I would not have any reader conclude that there is never an advantage in buying from a distance, for there often is a great advantage in it, but when buying stock of ordinary grades, if you have an opportunity to buy from a nearby breeder whose stock you can inspect, within the range of prices mentioned earlier in this lesson, the chances are that you will be better satisfied in the end than if you send the same amount of money for the same number of birds to a breeder at a distance. After one begins to be able to judge of the quality of his stock, to know where it is weak, and to know something about the characteristics of different stocks of the same variety, it will be his best policy to buy what he needs where he can get what suits him best, but a considerable part of the present buying away from home is of no benefit to anyone but the transportation companies. If every poultry keeper who had not a good reason for sending away for breeding stock and eggs for hatching would buy at home, sellers generally would sell as much as they do now, and the business would be on a much better basis.

There is another and a strong reason for the novice buying his first stock in his own locality if possible. Fowls, like all kinds of live stock, and like human beings too, are with few exceptions affected by change of climate. Nearly all fowls are unfavorably affected for a time — for a few weeks or months. After that some are likely to be better for the change, some worse, others not notably affected either way. On the whole the period of acclimatization is an unsettled period, and the beginner will almost invariably do better to work with acclimated stock.

But not all beginners can buy stock at home. There are still many localities in which thoroughbred stock is rare. Within a few years I have had a letter from a poultryman in a section where a show, at which he was an exhibitor, had been held annually for several years, asking for a description of White Wyandottes, one of our most popular varieties, and stating that they were unknown in that vicinity. So, while our popular varieties are quite well distributed, there are still many places where they are not to be had. The beginner located in such a place must go or send abroad for stock. If it is at all possible for him to go himself to yards where stock of the variety he wants is to be procured, it will pay him to do so. He can then see how the stock he buys compares with other specimens of the same stock, and have a better idea of its actual quality. Seeing the stock as it grows, he will also better appreciate the variations in it, and be better able to judge of the general quality of the stock he produces from it.

If it is impossible for him to inspect and select his stock, and he must buy by mail, he will still find it to his advantage to buy as near home as possible. By doing this he saves express, and, further, when the express charges do not constitute too large a proportion of the cost of the fowls delivered, he can return the fowls if not satisfactory.

Sometimes in making shipments a breeder will agree to pay return express on a lot of birds not found satisfactory, but in general the buyer pays express both ways. There is room for an argument as to what is fair in such cases. I think the best way for a buyer to look at it is this: —If he bought the fowls himself at the breeder's yards, whatever the distance from his home, he would not expect the breeder to pay his traveling expenses, or any part of them. He buys "on approval," because of the inconvenience or expense of going to see the fowls. He should then consider return express charges on unsatisfactory birds as an expense arising from his inability to inspect the stock before buying, and not an expense on account of the breeder's failure to send him stock that would suit, and therefore an expense for which not he but the breeder is responsible. If the buyer will look at the matter in that way he will enjoy more peace of mind than if he resents the payment of return express as an imposition for which the party who sent him the fowls is responsible.

In deciding from whom to order stock by mail, a beginner is necessarily very much in the dark. Many write to me for advice, information, and sometimes recommendations in this matter, saying that their resources are limited, and they cannot afford to make any mistakes. I can appreciate their position, and also approve the caution they display. At the same time I know of no way of avoiding "mistakes" in buying stock. I have been buying fowls, sometimes a good many and sometimes only one or two in a season, for sixteen years now, and I am somewhat acquainted with the quality of stock kept by a very large number of breeders, yet I cannot avoid "mistakes" of this kind. I don't think anyone can. The best I can do is to reduce the cost of mistakes in buying to the minimum by never buying more fowls than I actually need. That point we will take up again.

In deciding from whom to buy, give the preference (for reasons previously stated), to breeders nearest you. Write to as many of these as you wish, stating your wants, and asking for prices, and terms. Most breeders state their terms in a general way in their circulars, but not all are as explicit as is desirable. If you are going to buy stock on approval, (and it is the only way to buy if you cannot buy on inspection), you should have it distinctly understood before you send your order on what conditions the shipment is made. Some breeders make the statement that stock may be returned at their expense "if not as represented" in their circulars or correspondence. This is not enough. It is too easy to misinterpret both the buyer's statement of his wants and the breeder's statement of what he will send to fill the want. Make your order conditional on the stock being *satisfactory to you.* Have the seller's written agreement to that effect before you send him the money. If he will not agree to that don't buy of him. A man who will not make such an agreement may be perfectly square. Looking at the matter from his point of view he may think it not good policy to make sales that way. I think, however, that most breeders will without hesitation agree to such terms for shipments within reasonable distances.

How Much to Buy.

To make mistakes or misfortunes in buying of as little consequence as possible buy only for actual needs. I mean now your own actual needs. I think one of the greatest mistakes a beginner can make is to buy more stock at the start than he needs for himself, expecting that by selling eggs for hatching he is going to get back the additional money put into the stock, and more. The novice is wisest who lets the egg trade alone until he has his stock well estabished. One cannot be certain of the number of fowls or eggs that will supply him as many chicks for the season as he wants, but he can take the number that should do so if results are about average. Generally speaking, a breeding pen — that is, a male and four females — is enough to begin with. In more cases a trio will be better. Then if the fowls do not breed satisfactorily the outlay has not been heavy. There is, of course, a loss of time, but that cannot be avoided. Indeed, it is no uncommon thing for breeders of considerable experience to lose practically a whole season through the failure of their stock to breed satisfactorily.

Returning Unsatisfactory Stock.

The novice generally is not capable of judging the quality and value of the stock sent him accurately, nor is it possible to give him instructions that will enable him to arrive at an exact estimate; but there are some points that may be given that will help him to know some faults for which he should reject stock, and some reasons why he should sometimes hesitate to reject it on his own unfavorable impression of it.

Unless bought in the fall, birds that are decidedly immature and undeveloped should not be accepted. Breeders sometimes, in their eagerness to fill orders, send chickens whose quality is as yet uncertain, telling the buyer that they will grow into fine birds by the breeding season. If the buyer was advised before ordering that this class of stock would be sent him, he has no cause to complain, but such birds too often fail to develop into what the breeder said they would, because they are not, when received, what they should be at their age.

Sick, injured, or dead birds the buyer should refuse to accept from the express companies. This leaves the settlement of responsibility for damage where it belongs — between the shipper and the transportation company.

The mere fact that on inspection a pen of fowls fails to realize the anticipations one had been indulging of what he was to get, ought not to lead him to reject them as unsatisfactory. If the birds are apparently in good condition, take them out of the coop, put them in a pen by themselves, and observe them for a little while — at intervals for a few hours, or, at most, a day. Give them and the seller a chance. A reasonable time is allowed — and should be taken for inspection. Often birds just from a journey look quite different after a short rest. But don't keep the birds more than a day. If you are then still not satisfied with them, return them promptly. For whatever troubles may develop while the stock is in your hands, you are properly responsible. A buyer cannot, in fairness, keep stock for one or two or more weeks, and then ask the seller to make losses good.

Do not let a small fault, as you see it, decide you to return birds bought at ordinary prices. There are few specimens produced free from faults, and they are not for sale except at very high prices.

If one fowl in a lot seems to you decidedly superior to the rest, don't use that fowl as the standard of the value you should get at the price, and be dissatisfied with the others. It may be that in that one specimen the breeder has given you special value. It is not at all an uncommon thing for breeders anxious to extend their reputation to give customers some rather better birds than the prices warrant.

Don't allow any ideas you may have of distinctive marks indicating the absolute purity of stock of the variety you are buying, to lead you to reject or find fault with it. Many novices have a notion that pure bred fowls have certain distinctive features which invariably appear in well bred stock. This is not the case, and in rejecting a fowl for lack of such mark, one may reject a fowl that is especially valuable for other or general excellence.

Don't be too much influenced in your opinion of the stock you buy by the comments or judgment of others who see them. Consider all such on the basis of the speaker's actual knowledge of and experience with the kind of stock in question. On the whole, it is safe to give the seller and the stock the benefit of any small doubts you may have on points other than sickness, damage, or immaturity. Persons selling stock cannot afford to send out stock that will not suit, and in most cases where this is done the party doing it is at fault because of his lack of experience in selling, rather than because of any deliberate purpose to deceive or defraud.

Testing the Stock.

It is always safest to breed one's first pure bred stock as mated up by the breeder from whom it is purchased. The beginner should remember, however, that failure to get results that suit him may be due to his inexperience — to mismanagement of the stock. Even if his experience with inferior stock has been such that he is sure of his ability to grow good chicks, failure to get such from his new thoroughbred stock the first season should not lead him to discard it. In such cases I would advise buying a few more birds from another breeder, and breeding them separately, but continuing to breed the first stock bought, for it not infrequently happens that stock that did poorly the first season in a new place, does uncommonly well the next, and it is therefore poor policy to turn it off without a further trial.

LESSON X.

The Most Important Part of the Poultryman's Equipment.

BEFORE leaving the class of topics we have been discussing in the last few lessons, I want to consider more particularly some of the points that have had incidental mention in them. We have discussed a number of matters of prime importance to the beginner in poultry culture:—the possibilities of the business, the different branches of it, locations, markets, the kinds of stock, and various like questions, and I have tried to indicate how a venture in poultry keeping may be as free as possible from mistakes that use up the beginner's capital and often exhaust his enthusiasm.

But not only through these lessons, but constantly in correspondence with readers of the paper, I find that many express in varying degrees the feeling of the correspondent whose letters with my comments on them appear in this issue of the paper. They think it would be much easier for them to get along if I would explain and decide for them in everything, down to the last and least detail, and where I stop they imagine that there are other reasons for stopping than those that are given, and that my freedom of speech is restrained by consideration for other interests.

When in connection with this series of lessons, I organized a special class in poultry keeping, with the purpose of keeping more closely in touch with a number of poultry keepers, and following their work in its details, and advising them more definitely than is possible in the majority of cases, I found in connection with this class an obstacle I had not at all anticipated. By far the greater number of the class evidently failed to understand that the "Special Section" was in effect only an effort to give individual instruction, as far as it could be given under such conditions, to a limited number of readers of the paper. The questions they asked, and the comments they made soon made it clear that the general idea was that the special section students were to be given such an answer as they wished to any question they saw fit to ask.

At the time I was quite at a loss how to account for this, for I had tried in the announcements and in all correspondence relating to this plan, to make it entirely clear that I was not going to do these very things. I have since thought that this general misapprehension of the purpose of that plan was owing to the general feeling among novices that the greatest hindrance to quick and sure progress in poultry culture was in the impossibility of getting absolutely authoritative and reliable directions as to what to do in every instance where the poultryman is called upon to make a decision.

In a sense this is true:—but what to the novice seems to be the trouble is not the real trouble. His view of the situation is superficial. His difficulty is not that he can find no one to direct him, but that he does not know himself. He is undertaking to do things for which he has had no preparation, or inadequate preparation. The first and strongest reason for urging beginners in poultry culture to begin small and go very slowly, is that only in this way is it possible for those who learn by themselves (that is, who through the agency of books, papers, their own experience, and an occasional interchange of opinions with another poultry keeper, are self-taught) to be in any sense well prepared to make choice and decisions when necessary.

Where a man learns poultry keeping as he would a trade, occupation, or business in which a thorough training covering a period of years is regarded as essential to proficiency, he has, during all that period, frequent — perhaps daily — opportunities to see how others under various conditions make such choices as he would have to make if in business for himself. He can also see subsequently how events mark a course as wise or unwise, and through it all he hears from those interested in the matter discussion of all phases of the situation.

Now there are, as we all know, some men whom no amount of training, experience, and instruction will fit for positions of independent responsibility. They need always someone to oversee and direct their work, to plan for them and decide for them. And there are some men who, wherever you put them, learn very quickly, as compared with the average of their fellows. But I have never yet seen a man who in any position acquitted himself well who had not the particular kind of knowledge and skill needed for effective work in that position. He may not have acquired this knowledge and skill for the purpose of using it in this particular way, but it was knowledge and skill that could be applied, and when you hear of a man accomplishing wonderful things in a line new to him, you may be very sure either that his previous training adapted him to this work, or that the reports of what he is doing are exaggerated. In poultry culture the stories of phenomenal successes by novices are—I think without exception—greatly exaggerated, or refer to accidental successes never repeated. As I have said in one of the recent lessons, the plants that I *know* are successful have all been developed very slowly from small beginnings.

The attitude of the beginner who thinks he would get along all right if only someone would tell him just exactly what to do in each situation as it arises, after having outlined for him the general scheme upon which his poultry keeping is to be conducted, is practically that as he looks at it the poultry business is one that, with all conditions right or best, runs itself. Most beginners will unhesitatingly affirm that this is not their attitude at all, but in practice will still continue to show that that is exactly their position.

Mr. J. begins poultry keeping with White Plymouth Rocks, because in some way he has formed the opinion that they are the best fowls for his purpose. If his hens do not lay, and his neighbors' Buff Rocks lay well, he concludes that the fault is in the stock. It never occurs to him that it is in his inexperience. When he began poultry keeping he fed by the method that seemed best to him. Perhaps it was recommended by someone who used it very successfully. Not getting results by it, he casts about for another method. He takes that of the poultry keeper who, so far as he can learn, is most successful — just at that time. He succeeds no better by it, and tries another, with no better success. Perhaps he tries a dozen different ways of feeding before he begins to get a satisfactory egg yield. Then he thinks results are due to the particular feeding formula used. The fact usually is that out of his varied efforts in feeding he has developed judgment and skill in feeding, and the results he is getting are not due to special virtue in the feed in use, but to the cumulative knowledge and skill that have been acquired little by little.

There is another fact closely related to this which many do not appreciate. It requires a good deal of familiarity with a subject to enable one to grasp understandingly anything beyond the simplest and briefest statement of matters in it. Even in a series of simple and easily understood statements many will make no lasting impression on the mind of one not familiar with the matters treated. Some readers who thought the elementary lessons of last year very complete when they read them first, write me that in reading them over again in book form they find much more in them — much that escaped their attention in the first reading, though they read carefully. The printed matter is precisely the same. There is neither more nor less of it. The difference is in the reader. He knows more of the subject — perhaps more from other reading, but certainly much more by experience. This is his own knowledge — what he knows at first hand.

The first year that I was in the poultry business I bought breeding stock of several varieties, partly from and partly through an acquaintance in the business. I did not get the results I thought I should have had, and blamed it on the stock, and discarded it all and bought new stock elsewhere, meantime harboring the thought that my acquaintance's aid had been more of a hindrance than a help to me. By the time I was getting good results in breeding I had discovered that, in general, the trouble that first year was not in the stock so much as in my inexperience. Later I learned that the stock soon became what its owner made it; that the most essential things in applying methods, and in the management of fowls, were good judgment and a thorough knowledge of the fowls with which one worked, and the various articles used; that — in short — the poultryman was the determining factor in every poultry venture.

Nearly every beginner supposes that he appreciates this. He counts as an important part of his equipment for the business a liking for fowls, or for outdoor life, habits of industry and perseverance and intelligence. These are all necessary, but they become directly available and effective only as they are displayed in connection with practical personal knowledge of the details of poultry keeping. The acquisition of such knowledge requires time, because variety in experience is required to give one such a general working knowledge of any subject that he is ready to act promptly and quickly in any situation in which in his line of work he is likely to be placed, and the beginner in poultry keeping often finds himself in situations that are full of trouble for one who does not know just what to do and how to do it, while one who did have that knowledge would soon have matters right or on the way to mend with the least possible loss.

Again, in every poultry undertaking there is a very large element of chance. Opportunities come which only the poultry keeper who is qualified to use them can estimate at their true value and take and use to advantage. This is, of course, equally true in all lines of work, but I doubt whether there is any other line in which so few of those engaged in it have had a thorough training. The newness of poultry culture as a business, or on a business basis is largely responsible for this. As I have explained in earlier lessons, there are conditions in regard to training of poultry keepers which limit opportunities to get a thorough practical training. But because this is so it is not necessary that the poultryman who learns by keeping his own fowls should get his experience regardless of expense.

The waste and loss of capital and of inclination to continue poultry keeping due to efforts to learn it on a large scale, are appalling — the more so because they are so utterly unnecessary. There is absolutely no need of anyone losing any considerable amount of money in poultry keeping while establishing a business, if he follows the simple rule of increasing his stock no faster than he is sure — entirely sure of his ability to handle it to advantage. If when making his plans a poultry keeper will base them on what he has done, rather than on what someone else has done or what he hopes to do, there is little danger of his planning beyond his capacity to perform. To be safe in his venture he must observe the necessary relation between his own development in knowledge of poultry and the growth of his plant. He must remember that it is always possible to adjust operations to his ability, but not always possible to adjust his ability to the scale of operations on which he endeavors to work.

In such matters as the purchase of stock, of appliances, etc., the only way a poultry keeper can buy goods to suit is by buying and discarding until he gets what suits him, not being too hasty about discarding unsatisfactory stock or goods, but giving each a fair trial. What suits one does not suit another, and each has to suit himself. It is possible, and when it can be done it is advisable, for a while, for a beginner to rely to some extent on the advice and suggestions of others, though I believe the beginner gets along best and advances fastest who, while considering advice and suggestions, makes the decision for himself when the responsibility for it, and the consequences, are his, and no one else's. I believe it is better for the beginner who is learning by himself to put the responsibility for what he does with himself, and not with his advisor, for — after all — when he takes advice, and acts upon it, he usually makes a choice between the advices available, and so the responsibility really is his own.

He needs to consider that he is likely to make mistakes, to fail to fully understand instructions given him, to neglect to do some things that ought to be done, and, in a variety of ways, to be personally and directly responsible for things that go wrong. I do not urge this with the idea of making novices feel that others are always blameless if advice they give does not result satisfactorily. There are cases where information given is misleading, and the novice cannot know it until by repeated experience the truth is made clear to him. But the usual attitude of the beginner is to look anywhere and everywhere but to himself for the causes of his troubles, and this attitude is a greater handicap on his progress than anything else could be, for it keeps him always searching outside of himself for reasons for the things that go wrong, when the commonest source of trouble is his own inefficiency.

So I say that the most important factor in poultry culture is the poultry keeper. It is only as he makes himself a poultryman that one is able to get value out of good methods, good stock, a good location, and good general ability.

LESSON XI.

The External Parasites of Poultry.

IN his book, "Diseases of Poultry," Dr. Salmon gives a list of more than forty kinds of lice and mites which infest domestic fowls, — parasites which make poultry their special prey. If a poultryman takes pleasure in that kind of knowledge, and has the time at his disposal, there is no reason why he should not study these creatures, learn their names, study them microscopically, learn to distinguish between them, and in general become well informed on the subject. But practically all that is necessary is that he should know how to keep them out of his premises, both by preventing their increase and reducing their numbers when increased to the point where they become troublesome.

A few lice on a healthy fowl do no perceptible harm — so long as they are few in numbers. It is even asserted by some authorities that a few lice are beneficial rather than injurious, consuming dead cuticle and causing just enough irritation to prompt the fowl to dust itself — its way of bathing — regularly; this, however, is all theory, and has not been demonstrated. The evident facts are :—

That fowls are very rarely wholly free from lice.

That in limited numbers lice do little damage.

That when from any cause they become numerous they are a very serious pest.

Many writers on poultry topics preach constant preventive treatment as the only sure way of avoiding losses through the ravages of lice. It is a very common thing to see statements proclaiming lice as a constant menace to the health and profitableness of fowls, and urging the necessity of unceasing warfare against them.

Impressed by such statements many poultry keepers carry on a systematic treatment for lice which takes a great deal of time and labor, adding greatly to the drudgery of their routine work.

Such continuous preventive treatment is not necessary when fowls are kept under ordinary good conditions, with no conditions existing which are favorable to the increase of lice.

When I make this statement I do not wish any reader to misunderstand it. It will not do to assume that conditions are right, and therefore — in accordance with my statement — precautions against lice and treatment for lice are not needed.

Look at the matter from the other side. Find out — if you do not already know — to what extent lice affect your poultry, or would affect them under your management with special precautions against lice omitted, and then, if lice are troublesome try to find in what respect conditions supposed to be good, are favorable to the increase of vermin, and therefore bad.

The most frequent conditions favorable to the increase of lice, are debilitated stock, dark ill ventilated houses, and buildings that by reason of the abundance of conveniences in the fittings are difficult to keep clean, and furnish many nooks and corners to which the sunlight rarely penetrates.

HOW TO DESTROY RED MITES.

When lice once become established in a house in sufficient numbers to cause serious trouble, the most common reason for difficulty in exterminating them is lack of thoroughness in treatment. Often the treatment while of the right kind is done by piecemeal, and when repetitions of treatment are required the intervals between are allowed to be too long. I find that this is nearly always the case when complaint is made that usual remedies are not effective. I have often had letters from poultrymen who said that they found it impossible to rid the fowls and premises of lice, though the treatment — as they described it — left nothing to be desired.

It being out of the question to go back of their reports and ascertain the facts in any case, I several years ago concluded to let some of my own houses become badly infested with lice, reproducing as nearly as possible the conditions of the typical poultryman who found the lice too many for him.

So one season, beginning in the spring, I systematically neglected or omitted every usual operation which might prevent the increase of lice. By midsummer I had one house badly infested with red mites. It is worth noting in connection with the fact that under ordinary good conditions lice rarely become troublesome, that the mites did not appear in numbers that made their presence plain without close investigation, until the conditions became very bad. The droppings had been allowed to lie for months. Even then it was only after a period of nearly two weeks of very hot damp weather that the mites began to be noticeable. Then within another week the place became literally alive with them.

At the same time — in order to give the body lice a chance to develop — I omitted to make provision for the hens to dust themselves. So I had at once a flock of hens badly infested with lice, and their house alive with red mites. The ravages of the insects under such conditions began to be discernible almost at once. For the lice I did nothing whatever but provide dusting places as usual by spading up here and there in the yards a few square feet of ground. Had the hens been badly infested for a long time this would not have been sufficient. As it was, they made almost constant use of the dust baths for a few days, and soon had the lice reduced to normal numbers.

The red mites which prey on the fowls at night, and leave them during the day to hide in rough places or crevices about the roosts, are said to remain on the fowls during the day as well as night when very numerous, but I could find none on the hens in these houses by day, though they were in such numbers at the ends of the roosts that they could not begin to find places for concealment by day, and remained in a mass so great that a slight movement of the roost would make a great bloody smear of them.

The first thing done for these was to remove all roosts and nests from the house, taking out also the cleats of wood on which the ends of the roosts rested, which were screwed to the wall. Then I brushed down the walls thoroughly with a broom preliminary to whitewashing. In doing this, quantities of mites were brushed to the floor, and undoubtedly many of them worked back again, but I paid no attention at all to them.

I began treatment by applying to the mites on roosts and nests, taken out into the sun, various preparations, and carefully noting their action. I found kerosense effective, but did not think after trying chloro-naptholeum in water, applied to the mites and roosts with a brush, that kerosene was as economical. I did not feel like using it as freely as I did the water and C. N. Whitewash also was effective for all mites it reached, though not as quickly as the chloro-naptholeum preparation. I used some of this in quite a weak solution, pouring into a pail just enough to color the water up well, and with a brush threw and spattered it over the walls for some distance from the ends of the roosts, taking care to get it into joints and cracks as much as possible.

In one pen I used nothing but whitewash, putting it on the underside, edges, and ends of the roosts, as well as on the walls of the pen. In the other two pens I gave roosts, supports, and nests a free application of water and chloro-naptholeum, and then whitewashed the walls. Note that: — Every part of walls, roosts, and attachments was thoroughly treated at one time with some preparation destructive to mites.

The fowls roosted in the houses the same as usual that night; the whitewash not being yet quite dry, no special indications of mites were looked for until after the second night. Then enough mites that had escaped treatment had worked their way back to be quite conspicuous.

But here is a point for the poultryman who is combating mites to observe. The mites which escaped the first treatment were those which were most concealed, and, perhaps, some brushed to the floor in the preliminary sweeping which had worked their way back. My observation of mites on some pieces of board well covered with them to which I applied road dust freely was that many of them were killed by it. But these mites, having worked their way back to the fowls, and got a full feed, would not retreat to the inaccessible places in which they had escaped the treatment, but stopped in the first place that afforded a refuge, and after that one thorough application to all parts of the house I directed my attention exclusively to mites found on the roosts in the morning, working on the theory that it was easier to take time and gradually exterminate the mites as they remained on the roosts where they were easy to get at than to try to follow them to their furthest possible hiding places.

In the pen that had been treated wholly with whitewash, I would turn over the roost in the morning at intervals of two or three days, and whitewash it, thus killing all mites that happened to be on it. In the other pens I would do the same thing with the other preparation mentioned. At each treatment the number of mites found became less until within two weeks practically none could be found, and they made no more trouble until the next season. Then with the houses neglected for a little while they began to multiply, but were quickly checked by a repetition of the treatment. I have since used C. N. in water whenever traces of mites became numerous, but have never found it necessary to treat more than once a season. I suppose the better policy would be to make a thorough application, or possibly two or three in succession at the beginning of warm weather, and thus prevent their increasing to the troublesome points, but since my first experiment I have rather liked to have at least one opportunity during the season to demonstrate that the red mites were not so troublesome a proposition if one made a thorough job of the treatment.

The plain indication of the presence of these mites is found in their excrement, little grayish patches, like fly specks, on the roosts and adjacent parts which they traverse. When you see these specks you may know with certainty that the mites are there. If treatment is begun at once the mites may be practically exterminated with a few applications to the roosts and adjacent parts where their tracks are seen. Promptness is important as much on account of the loss of vitality to the fowls, and profit to the owner, as because of the greater difficulty of getting rid of the mites when their numbers have greatly increased.

With prompt and thorough treatment whenever signs of mites are observed, there is no need of regular weekly, or even monthly, applications to prevent their increase.

For lice on the bodies of fowls, as I have said, no treatment is necessary if the fowls are vigorous and have an opportunity to dust as they wish. Lack of inclination to dust is a symptom of lack of vitality. If a fowl showing such a symptom is found to be lousy, treatment for lice may be given a time or two, but unless the fowl then with such other treatment as is necessary on other accounts, begins to shows vitality enough to keep itself free from lice it is not likely to be worth doctoring.

It is when hens are incubating that they require treatment for lice by the attendant. They may not suffer if neglected. Indeed I have had many hens go through an entire period of incubation, bring off a good brood and rear every one with no treatment for lice at any time, but the conditions during incubation are so much more favorable to the increase of lice, and the annoyance of lice is so likely to cause hens to break eggs or leave their nests, that it is better to take regular precautions against lice.

Similarly with the chicks when hatched. I have this season a number of very thrifty broods from nests which I allowed to become badly infested with lice. These chicks have had absolutely no treatment for lice except the opportunity given them and the hen to dust as soon as the chicks were ready to come from the nests. Some of these broods, after being kept for a week or ten days where they could dust at will, were put in coops on the grass with no dust bath, but have shown no signs at all of lice. I do not advise this as a practice, but the experience is useful in showing how great a part the opportunity to take proper care of themselves plays in keeping stock free from lice; and I think it will nearly always be found that by making conditions unfavorable to lice and mites, and giving the fowls and chicks suitable opportunities to take care of themselves, systematic and regular precautions against these parasites may be reduced to two or three a year for the adult stock, and as many for each brood of young stock.

In this lesson I have mentioned only a very few insecticides, none but articles commonly used for other purposes, kept on hand in many homes, and easily obtainable everywhere. Chloronaptholeum is a special preparation, but as far as its use in the manner described is concerned, seems to me to be just the same as sulpho-napthol, which I have since used sometimes in the same way with the same results, and I suppose that there are other preparations of the same character which may be substituted. The point I want to impress is that a common article of household use, with water, everywhere abundant is an effective remedy for mites, and an inexpensive one. Hence there is no occasion for delay in treatment while waiting for some special insecticide ordered only when actually needed.

It would be possible to greatly extend this lesson, giving a long list of insecticides, both proprietary and home made, but of these I shall say nothing here except that if convenient any standard insecticide may be used, provided it is applicable for the special purpose for which an insecticide is wanted. For adult fowls use a powder, if it seems necessary to give them individual attention. For chicks use a powder, treating each brood as a whole. For lice and mites in coops and on fixtures, use a liquid, applying with spray, brush, or broom as most convenient. In every case be thorough and persistent. Let the intervals between treatments be short, a week or ten days for lice to which powder is applied, and two to four days for liquid appliances.

Don't keep up a constant warfare on possible parasites, not knowing whether they are there or not, but keep a sharp lookout for parasites and signs of parasites, and when necessary go after them aggressively.

LESSON XII.

Internal Parasites of Poultry.

PARASITIC worms — particularly intestinal worms — infest many fowls whose owners do not at all suspect their presence. By "infest" here I mean that the worms are present in sufficient numbers to be troublesome. Some good authorities say that intestinal worms, like lice on the skin and feathers of the fowl, are almost invariably present, but as long as they are not too numerous they make no trouble, and may even have some function of benefit to the fowl. Just what this is, or how it operates, I have never seen stated, nor so far as I have read on the subject have I seen any suggestion of usefulness for the gape worm which infests the æsophagus of the fowl.

The literature of this subject is not large. Salmon in the "Diseases of Poultry," devotes about thirty pages to worms, giving them, I think, more space than all other American poultry books combined. His material is drawn largely from European writers and investigators. Little original work in investigation of diseases of this class has been done in this country. Dr. Paige, of the Massachusetts Agricultural Experiment Station, has investigated several mysterious troubles in poultry yards in the state, and found worms causing the trouble, and has had a good many diseased birds at the station for observation and experiment. When I was last at the station he was testing remedies on a number of diseased specimens, and, as I understood, finding results too variable to warrant any general conclusions.

Salmon's treatment of worms, while the best accessible to poultrymen, is often far from satisfactory. He seems to write almost wholly after the European investigators of the subject, and is often too technical in descriptive statements.

Worms, when present in troublesome numbers, interfere seriously with the health of the fowl. Considering the conditions produced by them as diseases it is found that the symptoms are not marked until a rather acute stage, and even then are not so unique as to immediately identify them. The presence of the gape worm in the throat is most easily determined, yet "gapes" is commonly confounded with other troubles like gastritis, in which gas escaping through the mouth causes belching; acute lung troubles accompanied by labored breathing; or, indeed, any difficulty or distress in breathing. As — especially in small chicks — general weakness from any cause is apt to be accompanied by difficult breathing, it is readily seen that the possibilities of mistaking other things for gapes are quite unlimited. And, as a matter of fact, a great many reported cases of gapes are not gapes at all, and the general impression that "gapes" is a malady that annually ravages the crop of young chicks all over the country is a great big general mistake due to the fact stated above that a symptom which might be described as gaping accompanies other more common diseases.

With regard to intestinal worms we have just the opposite popular attitude. They are rarely suspected as the cause of trouble, and rarely discovered until diseased specimens or infested premises are examined by men with medical training. All in all, the detection and effective treatment of these parasites that live within the body of the fowl is one of the most puzzling propositions the poultrymen to whom it comes have to deal with.

What I can say on the subject is said from the rather peculiar standpoint editors sometimes attain. I have never seen a case of gapes, nor have I had any trouble with intestinal worms in my own fowls. I have frequently been able by reference to authorities on poultry diseases to indicate intestinal worms as a probable cause of troubles about which readers of the paper asked me, and in many such cases treatment for worms has seemed effective, furnishing reasonable grounds for concluding that worms caused the trouble, though I must say that in more than one case in which satisfactory result of treatment for worms was reported to me the correspondent had not been able to discover worms, and could only say that after applying the remedies conditions improved. So I am in the position, not of an authority on this or other diseases, but of a plain poultryman with perhaps a little more than average familiarity with both unprofessional statements of cases and the professional descriptions of diseases and prescriptions for the same.

The Gape Worm.

The disease known as the "gapes" takes its name from a small red round worm which attaches itself to the mucous membrane of the windpipe. The conspicuous symptom of the disease is the gaping which gives it its name. As has been said, gaping, while the characteristic symptom of this disease, is not peculiar to it, but is a symptom in several other troubles. So to make sure of the nature of the trouble, and of the proper treatment to apply, the windpipe should be examined for worms. If they cannot be detected by opening the mouth of the bird wide and looking into the passage, take a stiff feather, not too large, and having stripped the quill to leave only a little brush at the end of it, put it gently down the windpipe, turn once or twice, and then withdraw. If there are gape worms present some should be found adhering to the feather. If the worms are found, the only way to treat them effectively seems to be to operate on each chick separately with a feather, as just described, or with a looped horse hair, or a gape worm extractor made of fine wire. Anyone can make such an extractor for himself, using No. 30 wire. Take a piece about 12 or 14 inches long, double it, and then twist the two ends so that a loop just wide enough to go down the windpipe, and half to three-quarters of an inch long is left at one end, while the wires twisted together for the rest of their length, make the long handle for the instrument. When this is inserted in the windpipe, and turned around, the worms are cut loose, and what are not withdrawn with the wire will be coughed up by the chick. Several other remedies have been given. One that used to be very generally recommended was to put the chicks in a box, and cause them to inhale lime dust. This treatment seems to have survived on paper rather than in satisfactory practice, for though it seemed to have the indorsement of many writers I never could learn that it was effective.

When the disease is discovered on premises, give the affected birds the individual surgical treatment just described; then take precautions to prevent it in future. According to the best authorities, and also to the most observant poultrymen who have had to contend with it, the gape worm, (syngamus trachealis), is communicated to fowls through earth worms which they eat from ground on which chickens with the gapes have run. The eggs and embryos of the gape worm are scattered over the ground, some in the excrement and some coughed up by the sick birds. They may be taken by other chicks or fowls direct from the ground, but the common method of receiving them is believed to be through earth worms. It is said to have been shown conclusively that they are taken into the digestive tract of earth worms, and may be carried for some time there, and communicated to the chick by the worms it eats.

Sometimes the infested tract is small, and trouble may be avoided by fencing the chicks out of it. A lady in Pennsylvania wrote me some years ago that she found she had no trouble with her chicks if she kept them away from a particular spot in the garden.

A poultryman, some time ago, stated in one of our leading poultry journals that he raised chicks on infested ground by keeping them confined while small to pens or sheds, the ground under which had been treated with lime. In these enclosures the chicks could get no worms, and were free from gapes, while those outside soon became diseased. He says that he has found that if chicks are kept off an infested piece of ground for three years few gape worms will be left in it.

Where the gape worm is prevalent it is a most serious pest. It abounds most on wet heavy soils, that is, on soils least suited to poultry. On the sandy hills and knolls of New England, we have no trouble with it. It might get a foothold on some of our low, rather swampy spots, but very little poultry is kept in such places, the "sandy, well drained" locations having been favored more perhaps than their merits deserve.

When I was a boy in Illinois we used to hear much of the gapes, and from the character of much of the soil there, it is probable that many of the cases were genuine, but I never happened to come in contact with them. In Colorado, with its dry sandy soil, we had no gape worms. I am inclined to think that if a careful investigation of poultry and yards were made all over the country, and a map prepared to show the areas free from the gape worm, those in which it was found, but not generally as a serious pest, and those where it was very troublesome, poultrymen would be surprised at the small area actually badly infested. I mention this particularly because people so often, supposing they have a case of gapes, fail to prove it or find out what really is the trouble, and so allow some other serious trouble to develop to a stage where it is hard to deal with, when, if they had been more thorough in the first diagnosis they might have learned just what was wrong in time to treat the disease easily and successfully.

Intestinal Worms.

Salmon gives a list of forty-five parasitic worms found in the intestines of fowls and in the neighboring parts. Some of these are found only in one kind of fowls; others infest all kinds of domestic land and water fowls. These worms he groups as tape worms, round worms, flukes, and thorn headed worms, the most numerous in varieties and the most common in occurrence being the tape worms and round worms.

The general symptoms of worms in the intestines are the same. The kind of worm present can only be determined by finding worms in the droppings, or by post mortem, showing them fixed in the parts of the fowl. If the droppings as voided by the fowl before treatment show no traces of worms, a vermifuge may be given, and the fowl kept where its droppings are easily examined. It is not certain that no worms are present because none are evacuated. Some worms are very difficult to dislodge. But a dose of the remedy to endeavor to secure from the droppings confirmation of the suspicion of the presence of worms in the intestines of the fowl is the only way practicable for the poultry keeper, short of killing one or more fowls, and making a careful examination of the intestines.

The inexpert examination is, of course, especially liable to error, though often it does show the presence of worms. Wherever a state has its experiment station equipped for the examination of such animals and fowls, poultrymen in that state should communicate with the director or the veterinary at the station whenever a serious trouble arises which they do not understand. Write the experiment station authorities stating the case. If they find on correspondence that there seems to be a case requiring investigation they give directions how to proceed. In practically all of the eastern states the experiment stations are prepared to do work of this kind.

"The symptoms which indicate worms in the intestines," says Salmon, "are not very characteristic, but are such as would be expected from ill health due to any chronic disease. The birds become dull, weak, emaciated, isolate themselves, are indisposed to search for their food, are stiff in their walk, their plumage loses its brilliancy and becomes rough, they have diarrhea, and sometimes epileptiform attacks. In certain cases the symptoms develop rapidly, and the birds die as though from an acute disease. The most certain evidence of the nature of the trouble is the discovery in the intestines of large numbers of one or more species of worms * * * upon examination of birds from the flock which have died or have been killed."

For treatment Salmon recommends first hygienic preventive measures. Says he: "One of the most important of these measures is to move the fowls upon fresh ground every two or three years, or certainly in all cases where such parasites are frequently observed in the intestines of the birds. Another practical measure which may be adopted at the same time is to remove the excrement daily from the houses and destroy any parasites or their eggs which may be in it, by mixing with quick lime, or saturating it with a ten per cent solution of sulphuric acid. The acid is cheap, but requires that great care be used in diluting it, owing to danger of it splashing upon the clothing and flesh, and causing severe burns. It should always be poured slowly into the water for dilution, but on no account should water be poured into the acid, as it will cause explosions and splashing. When treating diseased birds these should always be isolated and confined, and their droppings should either be burned or treated with lime or sulphuric acid as just recommended. Without these hygienic measures, medical treatment can only be partially successful."

For medical treatment the same authority says:—" One of the best methods of treating tape worms in fowls is to mix in the feed a teaspoonful of powdered pomegranate root bark for every fifty head of birds. In treating a few birds at a time it is well to follow this medicine with a purgative dose of castor oil (two to three teaspoonfuls)." * * *

" For the treatment of the heterakis (round worm) Meguin recommends mixing santonine with the food given to the fowls. The powdered santonine may be incorporated in a cake, the dose being 7 or 8 grains for each bird. An efficient remedy is made by boiling an ounce each of male fern, tansy, and savory in a pint of water. The resulting liquid is mixed with flour, which is then made into pills and administered to the affected birds. * * * Oil of turpentine is an excellent remedy for all worms which inhabit the digestive canal. It may be given in the dose of one to three teaspoonfuls, and is best administered by forcing it through a small, flexible catheter that has been oiled and passed through the mouth and æsophagus to the crop. The medicine is less severe in its effects if diluted with an equal bulk of olive oil, but if it fails to destroy the parasites when so diluted it may be given pure."

The remedies given by Sanborn in "Farm-Poultry Doctor," are slightly different. He advises for round worm a two grain pill of santonine followed by a half-teaspoonful of castor oil. This to be given about an hour before feeding every other morning for a week. For tape worm he prescribes five drops of oil of male fern in one teaspoonful of sweet oil. This to be given before feeding in the morning, and the morning feed given about two hours after to be a warm mash of bran and milk containing for each bird one teaspoonful of castor oil.

The Last Resort.

When worms of any kind become so troublesome as to cause heavy losses it is probably the best policy to discontinue keeping poultry on the premises for a time proportionate to the violence of the epidemic and the general condition of the buildings and soil. On an old plant it might be advisable to keep no poultry for two or three years. On a new plant a thorough cleaning up and disinfecting preliminary to the introduction of new stock presumed to be free from the trouble should be sufficient.

To what extent losses of poultry are due to worms, it is not possible to say. Doubtless many epidemics of so-called cholera and dysentery are caused by worms, and the unfortunate poultry keeper never suspects the real cause of the trouble. It is for this reason that anyone engaged in poultry keeping who has heavy losses he cannot account for ought to try to have an expert examination of diseased fowls made. This will in most cases show where the trouble lies.

LESSON XIII.

External Characters of Poultry, and Their Values.

Introductory.

THE title of this lesson is borrowed from the chapter of the same title in Mr. Edward Brown's book, "Races of Domestic Poultry," to which I am indebted for the idea of attempting a complete discussion of the subject. The practical value of the matter seems to make it worth while to preserve such a treatment of it in form that will make it available for future distribution. Hence, I include it in the series of lessons, and adopt for it the method of treatment appropriate. Mr. Brown, in his treatment of the subject, seemed to me to limit the discussion to superficial characters, that is to those points of which fanciers make much, but which poultry keepers who are not fanciers generally regard with indifference if not with disdain, and to treat these characters solely with reference to their relation to the production of food values in eggs or meat. That may be the best present treatment of the subject for the British public for which primarily he writes, but in this country we have a very large class of poultry keepers whose aim it is to combine fancy and utility qualities, and in ever increasing degree we find poultrymen producing "fancy" and "practical" fowls from the same stock, the difference between them being a matter of individual selection, rather than of breeding.

Under such circumstances the proper disposition of a fowl becomes simply a matter of correctly classifying it, and using or disposing of it accordingly. It is not far from the truth to say that wrong principles of or errors in classifying fowls according to the purpose for which they should be used, cause by far the greater part of the dissatisfaction with results which buyers of eggs and stock for hatching manifest. To misunderstandings and misapprehensions of the points involved we must also credit the perennial discussion of "beauty and utility," which, in general, confuses more minds than it clears.

External Characters Enumerated.

Whether we consider a fowl as an individual whole, or consider it section by section and point by point, the various characteristics we see impress the eye by either form (including size) or color, or both. The impression in any particular may be favorable or unfavorable. Without attempting to make and maintain sharp distinctions as to their proper classification, I enumerate the following characters visible to and measured by the eye as having value or meaning beyond the mere fact that in themselves they please, or fail to please, the beholder:—

Size, shape, and carriage of body.
Size, shape, and carriage of head, (including head appurtenances — comb, wattles, ear lobes, crest, beard).
Size, shape, and carriage of wings and tail.
Size, shape, and carriage of legs and feet.
Color, quantity, and quality of plumage.
Color and texture of skin, both bare parts and those covered with feathers.

Even without any knowledge of ideas prevalent among poultrymen, a person would conclude after reading that list that many of these points were of special importance only under special circumstances, and that in a great many cases, if not generally, they were immaterial. And that would be a correct conclusion. But, to a very great extent, profitable modern poultry culture consists in the development and use of special features for special purposes, a proposition which brings us back again to the relations of "fancy" and "economic" qualities of "beauty" and "utility," and the possibility and advisability of producing from the same stock individual fowls to satisfy more than one special demand. The practical question, then, in each instance is whether the features the breeder would like to combine are compatible or necessarily antagonistic.

The Logical Fallacy.

In considering the possibility of producing for different purposes from the same stock, a great many fall into an error which apparently has its origin in the fact that most of us are by nature partisans, inclined to take extreme positions, and to regard those who will not go with us to our extreme as, therefore, at the opposite extreme, and the things which do not suit our purpose as, therefore, adapted to their needs, or produced by their methods. There is a very general disposition also to regard everything especially prized for some purpose we do not appreciate as on that account objectionable for purposes we seek, and this is responsible for no small part of the idiocies that enter into most discussions of the attitude of fanciers to practical poultrymen, and vice versa.

Such a statement as the above, when put in the form of a general proposition, always seems to overstate the case, but if the reader will note the literature of the never ending debate, and the frequent incidental references to points in controversy, I think he will soon agree that the application of a little simple logic to the situation would often show one the error of his ideas more forcibly than reams of argument in opposition.

When we prove a proposition false we do not necessarily thereby prove a contrary proposition true. Oftener we prove only that the point is immaterial, but that "only" may mean a great deal, may, in fact, be of greater importance than to have proved what was desired, for the more immaterial points we can establish the easier it becomes to cater to demands based upon such immaterial points, while still maintaining those points which we consider of first importance, and for our purposes essential.

Practical Value of Fancy Points.

Before taking up the detailed consideration of the points enumerated, let us consider briefly the practical value, the economic or cash value of fowls valuable for "fancy points." As we shall see as our investigation proceeds, some "fancy points" are of very substantial value, and fanciers are to be blamed not for preserving them, but for not giving more attention to them. But, in considering his own attitude toward points of no importance to him which the fanciers prize, the so-called practical poultryman should look at these points, not merely with reference to their value to his trade. He should consider their selling value to whoever appreciates them and wants to buy them.

While it is true, as often stated, that only a very small part of the poultry produced is or can be of the kind that commands high prices for fancy purposes, it does not therefore follow that it is better for a variety or breed of fowls, or better for most poultry keepers, or for the industry as a whole, that the regulation of standards and types should follow the ideas of the class containing the greatest number. The reasonable position to take in a matter of this kind is that, provided the qualities the mass of poultry keepers want are maintained, every point for which they care nothing that can be added is so much prospective gain, for it opens up possibilities of profit beyond what is possible in the class of fowls which suits the average keeper.

It is after we have reached this point that we come to the parting of the ways between "fancy" and "practical" ideas. It may, and often does happen, that a fowl, or the fowls of a certain stock generally possess certain characteristics or an all round high quality from the fancier's point of view, but have faults which make them far from desirable for economic purposes. Now as long as these are not faults which in time will also affect their "fancy" quality, the breeder who can sell them all at fancy prices can continue to be indifferent to those faults. Indeed it would be foolish for him to neglect the trade that pays good prices, and cater to that which halts at very moderate prices. He is working for the dollars and cents the beauty of his fowls will bring him, just as another is working for the dollars and cents he can get for their eggs and meat. The end of each is the same — dollars and cents. It is on the method of getting these from poultry that they differ. The question comes how far the same kinds, classes, breeds, and varieties of fowls can be maintained of the same uniform type and yet be well adapted to both purposes.

The study of the values of external characters of fowls should give results that will answer this question as well as indicate their relation to strictly consumptive uses.

Size.

It is evident from a comparison of fowls of sizes within the ordinary grades, that size has no necessary relation either to the prolificacy of the fowl, or the quality of its flesh, or to its vitality. When we come to a consideration of the extremes in size the conclusions are not so plain. In bantams, in which the smallest specimens are preferred, these small specimens are to some extent the result of inferior digestive and assimilative power, and in such cases it is reasonable to suppose that the functional inferiority which brought about the desired reduction in the size of the fowl would be a bar to productiveness — perhaps not so much to sexual activity as to high actual prolificacy. However, no study of that point has been made, and we must leave the question open. There have been instances of very good laying by bantams, and there does not seem to be reason to say that bantams in every way normal and with digestive power proportionately the same as that of medium sized or large fowls, should not lay as well. In fowls that are extremely large of their kind it sometimes appears that their great bulk resulted from failure to develop in some function — notably the sexual function, while digestive and assimilative power were equal to all requirements of the system. If at the age when sexual activity should manifest itself it fails to do so, and the usual quantities of food are consumed the result may be either excessive fat or growth, the former condition being, I think, by far the more common. Such conditions are abnormal. That very large fowls may be as prolific as any, has been seen again and again. The common opinion that large fowls are always coarse meated, I believe to be erroneous. The coarsest fibred poultry meat I have ever seen was on medium sized fowls, and I have seen as fine grained meat on very large fowls as on any. The shape of the fowl, especially the character of the bony structure, seem to me to have a much more intimate relation to texture of meat than has size.

The reasons given for some small and some large fowls being poor producers, or in some cases barren, explain how it might often appear that large fowls were lacking in productive capacity, when in reality the size resulted from some interference with the functional development of the fowl.

In small fowls again we can see how general constitutional weakness interfering with the full development of every function would bring about lack of size and lack of reproductive quality at the same time. This, however, would not be observed once in ten times for the other case, for very small fowls are apt to be marketed, while the best developed specimens are retained and often regarded as the most desirable from which to perpetuate the race.

But if we accept the reasons given we may admit that to some extent departure from the normal or average size may be objectionable. That the objection applies to actual size cannot be maintained in the face of results of comparisons between fowls of the same size from breeds in which the average size is different.

The final conclusion in regard to size is that in any long established breed the best results are likely to be obtained from specimens of the average size. To put it another way, a lot of average sized fowls of any breed taken as they run, would be much more likely to give good uniform results than an equal lot either below or above the average size. But in making and maintaining improvements in poultry, we always seek to go a little beyond average conditions or results, and when we find, as we sometimes do, specimens large or small of their kind with normal vitality, productiveness, and meat quality, we can, if we succeed in reproducing in its progeny the combinations in an individual, soon establish a stock of whatever type, including size, we desire.

IMPORTANCE OF SIZE — SHAPE OF BODY.

Apart from any possible or fancied relation between size and actual quality and productive capacity, the matter of size is sometimes of great importance. Certain sizes of fowls are in more general demand for certain purposes than others. Thus in our markets generally consumers call for fowls dressing four to five pounds each. This size is what is most in demand for ordinary roasters and for fowls. The demand for larger fowls is more limited, and many times the large fowl has to be sold at less per pound than the smaller one. Except in localities where there is a special demand for extra large roasters, it is no advantage to a poultryman producing for market to have his stock run larger than five pounds each. From the point of view of the fancier breeding Asiatics or breeding fowls of the American class generally up to standard weights, fowls dressing five to four pounds each are medium to small in size. Compared with the bulk of poultry marketed they are good sized to large fowls. Breeders of thoroughbred fowls divide quite sharply on the question of size, especially as to the desirability of maintaining full standard weights, or, perhaps, exceeding them. In the hands of most poultrymen all breeds deteriorate in size. Hence there is a very widespread demand for large fowls for breeding purposes to counteract this tendency. But this demand is not merely a demand for size. It calls for good size with general excellence and vigor. I think it may be truthfully said that great or excessive size in fowls that have no other claims to attention is not especially valued by one poultryman in a hundred.

Taking five to six pounds live weight as the most desirable weights for the poultry farmer, and therefore his standard weights, we may say that above these weights size is desirable sometimes in special markets, and quite generally in breeding birds, counteract the tendency the other way; while everything below these weights falls short of the standard for general demands, is a little less satisfactory for market, and to be used in breeding only with mates large enough to offset its deficiency.

Shape of Body.

Not as much now as some years ago, but yet a great deal, is said of the shape of the body of a fowl as an index of laying capacity. Like many other theories, the idea that a fowl with a long, deep, wedge shaped body is the best "machine" for the production of eggs, looks very plausible. But it has not stood either close observation or practical tests. Still there is a certain correlation between shape of body and producing capacity. There must be an appreciable fullness in the lines of the body to give suitable accommodation to well developed internal organs. The proportions may not be ideal. They may seem objectionable according to accepted standards of shape for the breed to which a fowl belongs, but if there is nothing in the shape suggestive of weakness or deformity, one type is as good as another for egg production as far as the evidence goes.

In table quality shape counts for a great deal. The most desirable carcass is that which carries the largest proportion of white meat. This meat is on the breast, body, and wings. It constitutes the muscular system for the wings, and there is therefore a very necessary correlation between good breast development and power of flight, (or perhaps I had better say capacity for flight). With power of flight, as a rule, we also find associated greater general activity. This is as true when we compare specimens of the same breeds as when we compare different breeds. I have often had occasion to note in handling Light Brahmas that those with well developed breasts were more energetic and active than the others, and consequently less liable to the ills which result so quickly from inactivity in heavy fowls. In such comparisons as this we must of course make allowances for the effects of condition and habit, but I think it will be found generally true that under the same conditions fowls with well developed breasts not only look more robust than those deficient in that section, but are inclined to be more alert and, as a class, are more free from disease, and have a longer productive life.

The apparent size of the breast does not, however, depend exclusively on wing power or capacity. Sometimes we find fowls which, when carved, yield more breast meat than their appearance indicated, because the bones of that region are contracted, reducing the space occupied by the heart and lungs, thus compressing those organs within the space they normally occupy, or by crowding them on others. Such specimens are usually those which a poultryman with an eye for good shape discards on general principles even before they have had time to show what excellence they might develop.

Conversely, we find many specimens much more deficient in breast meat than their appearance as they stand in life would indicate. In these the "trunk" is amply developed, but the capacity for flight is small. Naturally the most conspicuous instances are found in the larger breeds, but cases in plenty may be found everywhere. It is a fact sometimes lost sight of that in proportions of desirable meat such breeds as Polish, Hamburgs, and Leghorns excel so-called general purpose and table fowls. The objection to them for table purposes is their lack of size, and the precocity which makes the meat, especially in the males, hard and dry long before they have reached full growth.

Breadth and straightness of back are highly desirable attributes of shape in fowls. The apparent shape of the back, (more perhaps than of any other part, though all are subject to the same influence), depends much on its length, on the general carriage of the fowl, and on the quantity of the plumage. Close observation of this section in fowls will soon show the reader that it contains a great deal of "character." Apart from its outward expressiveness of strength and vitality which a good back gives in a fowl as in man and other animals, the straight wide back in the fowl indicates a normally formed body cavity in which the organs are in all probability constitutionally and mechanically perfect, while any irregularity or narrowness or deformity of the back indicates weakness of structure, possibly malformations of internal organs, and, quite certainly, limitations on the operation of the functions of those organs.

With good development of back and breast it is usual to find the abdominal region in keeping; but we do sometimes see specimens in which the development here is conspicuously lacking. The length and texture of the feathers on the abdomen have much to do with its appearance. If it is evident that the deficiency is not a superficial one, but that there is an actual lack of development here, I think it will be found that a female so formed is either a poor layer or a producer of small eggs, or that she is especially liable to trouble in passing eggs, all of which faults are associated with the fault in mechanical structure of the fowl. The posterior development which some regard as a sure "sign" of a good layer may be a help or a hindrance in the production of eggs, but it neither causes nor regulates egg production.

In males the abdomen is naturally not so well developed as in females, and has no particular significance as long as it does not suggest deformity. Males from a line in which the females are full in this section will sometimes be larger here than is consistent with a trim figure and pleasing carriage in a male, but such a fault belongs to the class known as "good faults."

Shape as the characteristic of a breed, is simply the type which those interested in the breed have decided upon as the standard. In some strains and stocks of fowls the breeder has succeeded in fixing shape to which his fowls conform with quite as much accuracy as to the standard for color, but in most stocks of all varieties variations in shape are common and numerous. It is only the occasional breeder and judge that give breed shape the consideration it should have. While the breeder of thoroughbred fowls for sale and competition will find some recognition of success in breeding fowls that are models in form, he will rarely find it fully appreciated unless combined with a superiority in superficial points which challenges attention. Then it is likely to get due recognition.

It is held by some that in each breed there is a certain shape which represents the highest utility development of the breed. According to the same authorities there is also for each breed a certain size with which the highest excellence is associated. It requires but a very limited observation of fowls to convince an unbiased mind that these views are both erroneous.

Carriage.

The carriage of a fowl, its habitual attitudes and movements, have significance first of all as an expression of its degree of vitality and health. When sick or tired the actions and attitudes of fowls of all breeds are singularly alike. When in health and good spirits different classes of fowls and sometimes to a lesser degree different breeds of the same class, exhibit peculiarities of carriage which are regarded as desirable characteristics of the breed, and which sometimes are a considerable factor in breed shape. But without strict regard to breed characteristics many specimens show a distinctive and pleasing carriage which commends them to a buyer, and often wins them a better place than they would otherwise get in competition.

Head Characters.

Fowls differ as much in size, shape, and expression of the head as in size and shape of body and color of plumage. Marked differences in head types may be found in the same variety, and even in birds of the same strain or stock. We are so accustomed to consider the head as a mere base for its appurtenances, beak, comb, earlobes, wattles, crest, beard, that it is only in a very few breeds that size and shape of the head proper are given particular consideration. The head least adorned with superfluities in flesh and feathers is the most expressive of quality, for in it the correlation with the other parts of the fowl is discernible, while a head profusely ornamented does not show for what it is. Undoubtedly the character is there, but the appendages are so much more prominent that it is not readily observed.

This lack of discernment of the character of the head proper is not, however, a serious matter. For as the head expresses with reasonable accuracy the character of some points of structure of body and of constitution, its appendages generally partake of the character of head and body, and by their greater conspicuousness show them even more unmistakably to those versed in their meanings. The comb, the wattles, the earlobes, the face, the beak, the eyes, all tell the shrewd observer something worth while for him to know. I would not be too positive in making a statement of this kind, but I think that though they may not always realize it, most critical judges of a fowl by external appearances are more influenced by the character of the head of the fowl than by the shape of the body, that the expression and impression on their eye of the head of the bird has an influence on their judgment of it as a whole which is not often changed as they pass the other sections in detail. To the average poultryman of tolerably keen perceptions, crests and beards have less meaning as expressing the general character of the fowl, but to the breeder of crested and bearded fowls who has studied them closely they mean much. Indeed it is in perception and appreciation of the correlation of the parts and qualities of fowls with details which ordinarily escape notice that the skillful breeder has his greatest advantage over others.

I do not think it can be shown that the size of combs, wattles, earlobes, crests, and beards has any special relation to any practical quality. Large combs are sometimes said to indicate laying capacity, but I have not found it so in individuals, nor will a comparison of breeds differing in size of comb substantiate this idea.

The shape of the comb we must consider in two ways: First, as to the kind— single, rose, pea, leaf. It is often said that rose and pea combed fowls stand cold better than those with single combs. This is but a half truth. It depends as much on the size as on the shape of the comb, and we have to consider also the development of the wattles, with relation to susceptibility to cold and frost. A very large fleshy rose comb that has no spike, is lumpy and only a "rose comb" because it is not anything else, is often associated with very long, pendulous wattles which freeze as quickly as the wattles of a large single combed fowl. A rose comb of the type generally preferred in Wyandottes is usually associated with a medium sized wattle not readily affected by cold. A single combed fowl having wattles of the same size generally has a comb low enough not to be affected by frost at any temperature that would not affect the wattles. We must count size of comb then as a point indicating in a general way the capacity of a breed of fowls to resist cold.

When it comes to the matter of individual resistance to cold, the comb, unless quite inconspicuous, becomes a very accurate indicator. Observe your fowls on any raw cold day, and see how some combs are bright as usual, others slightly discolored, and others quite blue or purple. The discoloration of some combs may indicate only constitutional susceptibility to cold, or it may mean that the fowl is at the time out of condition and therefore unusually susceptible to climatic influences.

The texture of the comb is also a point of importance. The fancier prizes fine texture in the comb for itself; the poultry grower because he considers it an indication of fineness of fiber in the meat of the fowl. The carriage of the comb in single combed fowls conveys impressions which probably are misleading, though I would not care to dogmatize on this point. A comb that is erect and smooth looks strong and gives the fowl an appearance of strength. A comb that bulges, bends, lops or wrinkles suggests weakness, just as any lack of symmetry does, but I doubt whether any ratio dependent upon such variations could be established.

Considering the shape of the comb and wattles as compared with others of the same kind: It is found that there is a type of unsymmetrical comb quite common in all fowls, and especially conspicuous in single combed fowls, which gives an expression in the fowl corresponding to that made by a human profile with retreating forehead and chin. With such a comb is usually associated a rather long narrow head and a bill that without being abnormally long, and giving the fowl a "peaked" look, yet conveys an impression of lack of force and stamina. It is commonly held that such fowls are lacking in sexual vigor. I think this opinion requires modification. My observation of such fowls suggests that the reproductive faculties are as active in them as in others, but that their operation tends to exhaust the vitality of the fowl more rapidly, that it is a lack of general stamina rather than of sexual capacity. The precise understanding of it, however, is not a matter worth quarreling over. It is a quite well assumed fact that such birds are not desirable breeders, and the hens not likely to be as persistent layers as those more symmetrical in these points.

To the shape of the bill fanciers attach considerable importance. In nearly all breeds a strong, sometimes stout, well curved or moderately curved bill is required. Such a bill looks better than one that falls short of the specifications. Market poultrymen find the strong, stout, well curved bill the mark of a good feeder. Experimenters in special fattening methods say birds with such bills are much better subjects for fattening.

The color of the bill is a point of importance to the fancier. He requires a bill of a certain color in each breed. To the market poultryman the color of the bill is of importance only as it indicates the color of the skin. The bill is almost invariably of the color of the legs, and generally the skin of the fowl corresponds both as to kind and quality of color. Thus in bill and legs one may judge the color of the skin of a fowl without examining under the feathers.

To the color and expression of the eye breeders attach great importance—far more than a careful comparison of results has ever showed me. It is held that a red eye indicates vigor and vitality, especially sexual vitality. It certainly gives an impression of boldness and strength to a bird, but I question whether this impression has any better basis than general opinion. We in this country also think white or flesh colored bills and legs make a fowl look weak as compared with one in which bill and legs are a good yellow, but the facts about the breeds do not substantiate this view.

The Wings and Tail.

The size and shape of the wings of a fowl would naturally be expected to be in proportion to the development of the breast, which is largely the development of the muscles which move the wings. In general the proportion is probably maintained. This we may conclude from a comparison of breeds. How closely the correlation of size of wings and development of breast is maintained in individuals, I do not think has ever been made the subject of inquiry.

We look usually for a similar degree of development in the feathers of the wings and tail of the fowl, and such probably exists, according to a natural law, except where, by artificial selection, different degrees of development have been brought about. With general similarity of development there is always much individual variation. This probably has no particular meaning. Indeed, if one were to attempt to make the supposed meanings or values of different external characters harmonize with each other he soon arrives at a situation where he must either admit that some things he felt very sure indicated certain qualities are really immaterial with respect to those qualities, and their simultaneous occurrence merely a coincidence, or, that the whole subject is so complex that he cannot follow it in detail. For instance: We may note in general that a fowl with large tail has large wings. The tail of the male bird is a sexual character, and its development is with some reason supposed to bear some relation to his sexual development. That it is a measure of such development I would not affirm, though it is well known that males with "hen tails," the Standard type in Sebright bantams, are likely to be sterile or quite so. For a male to be well furnished with distinctive male plumage, neck hackle, saddle, hangers, and tail with abundant coverts is generally supposed to indicate virility.

There is as we have seen, a general correlation of the size of the wings to the development of the breast. Now, unless we assume that there is no correlation, or no meaning in correlation of the stiff or main tail feathers with the soft feathers of the tail which cover them, we find, reasoning from one conclusion to another, that a well developed breast is an indication of sexual activity. And if we pursue this same course of reasoning from section to section throughout I think we will finally resolve the whole matter into the simple general proposition that the fowl that is best developed physically, that is, with the best all round development, should be the most productive fowl and the most reliable breeder. Such a proposition looks very reasonable, but every experienced poultryman and breeder knows that though partly true it is true with so many modifications and limitations that it does not furnish the simple rule it would if absolutely correct.

The carriage of wings and tail we do not readily dissociate from the general carriage of the fowl, for its habit with regard to them is what makes fully fifty per cent of its distinctive carriage. Fanciers find that bad faults in carriage of these parts, whether due to malformations or to laziness and general lack of liveliness, are very persistent in the progeny of fowls. Not only so, but they are distinctly a handicap to a fowl in the exhibition room, and also contribute to an unfavorable first impression which makes them less salable than birds that are full of life and style.

Any departure from the usual habit of carrying tail and wings indicates a degree of exhaustion or perhaps the presence of disease. A fowl that is weak and debilitated carries its wings and tail as if they were burdens and incumbrances, dragging it down. A fowl in the full vigor of health carries wings and tail so easily and naturally as to attract no special attention to them as parts of itself, or uses them so handily that they add to the impression of beauty, strength, and gracefulness which it makes.

The Legs and Feet.

In the shank and foot of the fowl we have the most reliable external sign of some things of importance. It is the only part of the frame of the fowl not so covered with feathers that we cannot judge by sight of the general character of the skeleton or frame work of the fowl. The shank, and especially the clean, unfeathered shank, indicates very accurately the character of the skeleton of the fowl. If it is fine with small, neat joints, that same structure will be found throughout, and in such a fowl there will appear to be a larger proportion of edible meat and less waste in bone, tendons, and cartilage. The common belief is that this not only appears to be, but is actually the case. Mr. Brown, in the chapter of his book to which I have frequently made reference, quotes M. Abozine as stating at the poultry conference at St. Petersburg some years ago, that "on examination of the skeletons of a large number of fowls he always found that the relative weight of the dried skeletons to that of the entire living bird and its edible parts is the same for all breeds, and equivalent to six per cent." Mr. Brown quotes this statement as one calling for further inquiry. I would not deny it in the face of what purports to be a careful report, nor would I indorse it without more extended investigation. If it is correct there is no variation in proportions of weight of skeleton to flesh, and some of our ideas about producing table poultry carrying the largest possible proportion of edible meat, seem to be wrong.

However, this theory, though it may, if proved, contradict the old dictum that in a fine boned fowl the skeleton is produced at less cost, does not necessarily interfere with the idea that fineness of bone is associated as a rule with fineness of flesh. This, I think, is the general and apparently well founded belief of most of those who handle poultry. Prof. W. R. Graham, in a recent lecture called attention to the texture of the skin of the shank of a fowl as always indicating the texture and quality of the flesh as well as of the skin covered with feathers. This point I had noted in a number of cases, but not enough to enable me to be as positive of its general truth as he was. Knowing how thoroughly he has gone into the subject of table poultry, I think his judgment on this point worth acceptance.

The carriage of the feet and legs — the fowl's way of standing and walking — go far toward the making of the impression it presents. To some extent the position of the legs and feet are dependent upon the general structure of which they are a part, yet often peculiarities in regard to them are plainly local, and in their contribu' on to the carriage of the bird they express its spirit quite as faithfully and conspicuously as do some of the superior sections. I do not know that there is anything indicated in the legs and feet of a fowl that cannot be known without special reference to them. Their indications of structure and of quality of flesh are discernible also in the head and comb. The testimony of different parts of the fowl to the same facts, however, is cumulative, and ought to make the requisite impression on the person seeking for evidence of quality, or lack of it, much more forcible than if confined to a single section. Further, all do not value the same sections alike, and evidence of a fault appearing in one section might make little impression on one mind, while the evidence of the same fault in another section would make a very strong impression. So, also, with respect to excellence. Some require one sign, some another, of the same quality. Whoever depends on a single point is apt to find his judgment on it occasionally leading him astray, while with two or three points of veiw, if his udgment misses on one he is likely to be set right on another. Thus I have known men who saw in a peaked looking beak and head only evidence of "fineness," but when they looked at the unsymmetrical, rough, or shriveled looking shanks and toes of the same specimen, saw at once that it was weakness.

The condition of the legs is generally considered a reliable index of the age of a fowl, but it is as often misleading. To mistake the legs of an immature fowl for those of a mature fowl, if both were in good condition, wou'd be quite impossible to most experienced poultrymen, but as soon as the legs and toes begin to be out of condition, either because of external conditions or because of the condition of the fowl, judgment by them becomes all guessing, while in the case of mature fowls age alone may make little difference in the appearance of the leg of the yearling and the bird of two, three, or four years.

In young chickens the leg and foot often furnish the first indication of trouble. Especially is this the case with chickens raised artificially. Improper temperatures and lack of ventilation soon show their effects on the feet, which seem to wither away, while the chick that finds conditions right has a smooth skinned, well rounded, sturdy looking pair of legs under it.

Color, Quantity, and Quality of Plumage.

The color of a fowl in no way influences any other quality, though color defects are sometimes (perhaps rightly) held to indicate a degree or period of weakness. Thus in black fowls if a feather containing some white or gray, when plucked, is replaced by a solid black feather, the presumption is that the first feather was not perfect because of some lack of perfect condition in the fowl. But it would be hard to show by comparison of black fowls that those with a trifle of white in the plumage were in any way, except in this variation from Standard color, inferior to either the solid black birds or to those which, from excess of coloring pigment, showed purple bars in the black. The actual difference is too trifling — even if it does exist — to have an influence marked enough to be noticeable.

Superficially, however, and considered with reference to the demand for certain colors and markings in exhibition fowls, and with reference to such points as ease of dressing and the better appearance of fowls of certain colors when dressed, color is of great importance. With the fancier excellence in color compensates for many serious faults in other matters, and a fowl remarkable for color will sell or win regardless of other faults where a fowl poor in color would not be considered for its other merits. It is the fowl of this character which the "utility" poultry keeper should buy of the fancier — a cull for superficial faults in no way affecting any substantial quality. This is a point both breeder and buyer should keep constantly in mind. The utility poultryman wants not any cull from the fancier's yard, but the bird which is a cull from the fancier's point of vision, and at the same time not a cull from his own.

The history of the popularity of varieties of poultry indicates that even those who do not breed for high excellence in color and markings are not long satisfied with fowls that, as they run with ordinary selection in breeding, show too great diversity of color. This accounts in part for the steadily increasing preference for white fowls — the fact that the white fowl is at any age easy to dress clean, the pin feathers not being full of pigment which, if it exudes, stains the carcass, and if it remains in the quill disfigures it.

The quantity of plumage, profuse or scanty, is advantageous or otherwise according to conditions. The Asiatic with abundant plumage will stand cold and exposure better than any other fowl, but in a long experience with Asiatics I have found that in extremely hot periods a much larger proportion of the adult fowls will die simply from the effects of heat than of any other fowls.

The young stock, not often being in full plumage in summer, is not often so affected. Fowls with short or scant plumage stand hot weather best, and are most susceptible to cold and to such temperature changes as are common in fall, winter, and spring. There seems to be a close connection between abundance of feathering and foot feathering. Not that the two points are not separable, but in general in a feather legged breed the volume of the foot feathering is in proportion to the length and abundance of the plumage of the body. It is notable, also, that an excess of plumage on the body and feet of the fowl is often, (if not generally) associated with a shortening of the feathers of the wings and tail, and (whether there is any necessary relation of these facts, I do not know) twisted wing feathers and wing feathers narrow almost to deformity, are far more common in heavily feathered fowls than in those in which the development of feathers is more moderate.

The quality of the plumage, like the texture of the skin of the fowl, is an index of its physical condition, and varies with it. This is a point to which little attention is given by any class of poultrymen.

Color and Texture of Skin.

Some reference has already been made to the texture of the skin of the legs in relation to the texture of the flesh. The fancier, as a fancier, takes no specific account of the skin of the fowl that is covered with feathers, but if the skin of the exposed parts, the head, legs, and feet, is what it should be the skin that is concealed is not at all likely to be wrong, for there is a natural correlation of texture and quality of the skin which extends to all parts.

The skin of the exposed parts not being as desired or required, does not necessarily mean that the skin of the body is not right. There are numerous diseases and some conditions which affect the exposed skin, but may not extend further. The bleaching of the skin of the legs and feet of fowls which run on very dry earth or ashes may be quite marked, yet the skin of the body be of good color. Dust and dirt may disfigure the comb of the fowl so much as to give it a faded or unhealthy looking color, yet the protected skin show no such effects. The texture of the skin of the legs is of great importance in the matter of resisting disease and conditions injurious to the skin. Fine grained, firm skin and scales resist to a remarkable degree the action of soils, ashes, etc., and also of the insects which cause the loathsome disease known as scaly leg.

In general it will be noted that a fine soft oily skin is associated with fine quality, firmness, and glossiness in the plumage. These conditions or characteristics are to some degree dependent upon the constitution and general health and condition of the fowl, but it seems quite clear also that this quality contributes much to the health and good condition of the fowl. It may be a question of opinion which is the cause and which the effect in such matters, but the value of the external characteristics and their meaning are apparent.

In most American markets the demand is for poultry with yellow skin and legs. In this respect the color has an actual cash value for table purposes, just as color of plumage has for exhibition. Of course in the consideration of single specimens, the value of color of plumage may be represented by many dollars, while color of the skin of a table fowl increases its selling value only a few cents, but the demand for yellow skin in table poultry is general and insistent, while preferences for colors of plumage are various and changeable.

In Conclusion.

We have seen as we have discussed these various characteristics that in the majority of instances where a character has special value, it is because it meets an artificial demand. This is as true of "practical" as of "fancy" points. If demands change we change our fowls or our styles, sizes or types of fowls accordingly. As we have, even with the most careful breeding and skillful growing, considerable variety in each year's product, the profitable disposition of

It depends on selecting for each demand to which one caters such fowls as will best meet that demand and would not meet a better paying demand. Not all good qualities can be combined in equal degree in the same fowl, but by proper selection of breeding stock year by year a breeder can have stock in which he combines with good market and laying qualities the possibility of great excellence from a fancier's standpoint. If he does this he has always three classes of prospective customers—fanciers, farmers, and poultrymen who wish fowls which they are not ashamed to have those who know good fowls see, which will at the same time give a good account of themselves at the nests and on the table. All these can be supplied from the same flock, if the needs of the two extreme classes are always jointly considered in breeding. If either is neglected the stock is quickly brought to the point where only one class can be satisfactorily supplied. No matter which class this is the possibilities of immediate profit, and what is more important, the opportunity to extend one's business are greatly curtailed.

Breeders should distinguish between essential and non-essential points, work for every point essential to any profitable demand, and despise no non-essential because it is immaterial to them personally.

LESSON XIV.

Poultry Nomenclature and Abbreviations.

BEGINNERS in poultry keeping, interested — as many are — in everything pertaining to poultry culture, and regarding every breed they hear of as a possible favorite, are often puzzled by the use in poultry literature of abbreviations of names of varieties. The habit of abbreviation is so universal that poultrymen need make no apologies for it. At the same time, there are so many varieties of poultry, and so many varieties and sub-varieties have names that to one who has an idea of the breed constitute a very good description of them, that it is worth while for a novice to inform himself in regard to the abbreviations used.

As nearly everywhere, abbreviations of poultry names are not made to follow one consistent system. The method of giving names, while to a considerable degree systematic, is not wholly so. What, considering other varieties and names *should be* the name of a variety, is not always the name given it. An abbreviation of a part of the name of a variety commonly used for it, and equally applicable to the name of a second variety, may never be used in the name of the second. Hence in making abbreviations one must consider what is common usage much more than what would be logically correct. This is a point the reader who occasionally writes either to give or to ask information about fowls, should keep in mind. An abbreviation may be admissible, but if people are not in the habit of using it many will not take the trouble to be sure they know just what is meant, and sometimes those who do try to make sure of what is said have difficulty in doing so. Indeed the use of abbreviations by those not familiar with most of the abbreviations of names in common use often use an abbreviation which does not clearly distinguish the fowls to which they apply it because equally applicable to another variety. The most troublesome cases of this kind are where an abbreviation in common use for years is made ambiguous by the advent of a new variety having a name with the same initial. The boom in buff varieties caused more trouble in that way than any other recent occurrence.

In Plymouth Rocks we have three varieties — Barred, White, and Buff. The original Plymouth Rock was the Barred variety, and for a long time Plymouth Rock meant nothing but Barred Plymouth Rock. Even now the club devoted to Barred Plymouth Rocks is "The American Plymouth Rock Club." With the introduction of the White variety it became necessary in speaking of Plymouth Rocks to differentiate. Even before this it had been common to speak of the Barred Plymouth Rocks simply as Rocks. Now in addition to the abbreviations B. P. Rock and W. P. Rock, we had B. Rock and W. Rock for Barred Rock and White Rock. Occasionally someone would use the term Barred Plymouth or White Plymouth, but such usage was rare. With the coming of the Buff Plymouth Rock an element of doubt was introduced into the use of the initial B. as an abbreviation for names of varieties of Plymouth Rocks. B. might stand for either Barred or Buff. Though more than ten years have passed since then the plain B. is still often used, especially in connection with the Barred variety. Some poultrymen and writers make a point of doing this on the ground that the Barred had preempted the use of that abbreviation. Enough, however, will not look at it that way to make it generally uncertain what is meant by a B. Rock, or B. P. Rock. I have sometimes used Br. for Barred and Bf. for Buff — a usage which looks all right when only Rocks are considered, but as Br. is becoming somewhat generally used for Brown Leghorns and for the word Brown when occurring in the name of a variety of Games or Game Bantams, Bd. for Barred would be preferable.

The breed name Wyandottes is very commonly shortened to 'Dottes, and abbreviated to "Wy." The original Wyandottes were the Silvers, — a laced variety. In the American "Standard of Perfection" they are still called Silver Wyandottes, and the abbreviations S. for Silver, and S. Wy. for Silver Wyandotte are common. Since there has been also a Silver Penciled Wyandotte, there has been an increasingly general use of the more complete descriptive name Silver Laced Wyandotte, abbreviated sometimes to S. L. Wyandotte, or S. L. Wy. For the Golden Laced variety the abbreviations have been the same with the substitution of G. for S. For Whites the initial W. with the abbreviation Wy., is very common. Buff and Black having the same initial, the abbreviations may be Bf. and Bl., though there is no established usage. Partridge Wyandottes are properly described as Golden Penciled Wyandottes, though the other is the "official" name of the breed. Partridge may be abbreviated to P. or to Part., which is more suggestive, but I think you will find the name used in full many times oftener than you will find it abbreviated. Silver Penciled Wyandotte may be abbreviated to S. P. Wy., or Wyandotte. It is quite common to call them especially Penciled Wyandottes, but that leads already to the habit of speaking also of the Partridge as Penciled Wyandottes, and I frequently get communications speaking of Penciled Wyandottes, which leave me in doubt as to which is meant. For Columbian Wyandotte Col. Wy., or Wyandotte, seems right, and is occasionally used, though Columbian without the breed name seems to gain favor.

For Javas and Dominiques no abbreviations are in use. Rhode Island Red is commonly shortened to R. I. Red, or simple Red. The two varieties being distinguished by the shapes of the comb, as Single Combed or Rose Combed, it is common to abbreviate either to S. C. R. I. Red, and R. C. R. I. Red, or still further to S. C. Red, and R. C. Red.

The varieties of Brahmas — Light and Dark — have variety names which indicate their colors as compared with each other. These descriptive terms are abbreviated sometimes to Lt. or L. for Light, and Dk. or D. for Dark, but the abbreviations are not as much used as the full names.

In Cochins we have the varieties Buff, Black, White, and Partridge. W. Cochin, and P. Cochin, for the two latter cannot be misunderstood, but to be sure of the others we must add a letter to the B.

Langshans have but two varieties — Black and White, the names often abbreviated to B. Langshan, and W. Langshan.

In Leghorn names, the Brown for years was given a monopoly of the use of the initial B., though the Black might claim it, and there was little if any trouble, because Browns were common and Blacks very rare. I suppose there are in this country today several thousand persons who have Brown Leghorns to every one who has Blacks. But when the Buffs came with some promise of popularity, it had to be Br. and Bf., or else use the words unabbreviated — which has perhaps been the more common practice. The Browns, Whites, and Buffs being subdivided according to the shape of the comb, we have the abbreviations S. C. and R. C., which were explained above in connection with the R. I. Red. In all of these varieties when there is no reference to the comb the Single Combed variety is usually meant. Silver Duckwing is abbreviated to S. D.

The two Single Combed varieties of the Minorca are the Black and White, and the usual abbreviations for these color names used without reference to the shape of comb will almost always mean the S. C. Minorcas. In writing of the R. C. varieties of the same colors the abbreviations of the full name should always be used.

For the Spanish, the full descriptive name (or its abbreviation) of White Faced Black Spanish is very generally used, though there is no possibility of error in speaking of Spanish, there being only the one variety. For Andalusians and Anconas we have no abbreviations.

For Dorkings the Silver Gray is quite generally abbreviated to Silver Dorkings, Gray Dorkings, or S. G. Dorkings, but the White Dorkings and the Colored Dorkings names are not often abbreviated. For Redcap no abbreviation is used.

For the Orpington variety names we have the usual abbreviations for the colors of the varieties, White, Black, and Buff, — and for the subdivisions of these according to shape of comb. Spangled and Jubilee Orpington are not abbreviated.

Of the Polish, the White Crested Black is the only name generally abbreviated. This is reduced to W. C. B. Polish. There is a Buff Laced Polish, and there are both Bearded and Non-Bearded divisions of the Golden, Silver, and White varieties, but these names, perhaps because of the infrequency of occasion for using them, are usually written in full.

For Hamburgs, the names White and Black, are but rarely abbreviated; the names Golden Spangled, Silver Spangled, Golden Penciled, and Silver Penciled, are quite generally reduced to the initials giving G. S., S. S., G. P., and S. P. Hamburgs.

No shortening or abbreviation of any kind obtains in our use of the names of the French breeds — Houdans, Crevecœurs, and La Fleche.

Abbreviations of names of Game and Game Bantam varieties cause most inquiry. A B. B. R. G. Bant. is a Black Breasted Red Game Bantam. Omit the Bant., and you have the abbreviation for Black Breasted Red Game. In both the large and the small fowls there is also a Brown Breasted Red, known as a Brown Red. The other is sometimes called simply a Black Red. As far as proper discrimination in abbreviation of these names is used, the usual way is to write B. B. R. for the Black Breasted Red, and B. R. for the Brown Red. Golden and Silver Duckwing Games and Game Bantam are abbreviated to G. D. and S. D., etc. For Birchen, Black, and White, abbreviations are seldom used. Red Pyle is abbreviated to R. P.

In the "Standard of Perfection" the word Game is dropped from the names of Cornish Indian and White Indian Games, and they are known simply as Indians, Cornish, and White.

The precedent seems to have found little favor, and the old usage quite generally continues with the abbreviations C. I. Games and W. I. Games. For Malays and Sumatras no abbreviations are used.

Golden Sebright and Silver Sebright Bantams may be abbreviated either as to one or both descriptive names. Rose Comb Bantams are of two colors — Black, and White. It is customary to use the abbreviation R. C. (there is no S. C. variety of the breed), but not to abbreviate the color term. In fact, it is not uncommon to see it written R. C. Blacks, or R. C. Whites, with particular reference to their being Bantams. For Booted White Bantams no abbreviation is commonly used, except that as in all varieties of Bantams it is quite customary in writing the name to shorten Bantam to Bant., even when the rest of the name is written in full. With those who abbreviate intelligently it may be said to be the rule—perhaps not specifically framed in the mind, yet still generally observed in practice to abbreviate, to cut short, as soon as the meaning is clear. In names of Brahma, Cochin, and the three varieties of Polish Bantams, we have the same sets of abbreviations as for the larger varieties of the same name, always, of course, with the name Bantam or its abbreviation added.

Of Japanese Bantams there are three standard varieties, — Black Tailed, White, and Black. The first of these is more fully and correctly described as a Black Tailed White Japanese Bantam. This gives us a very long name — for a very small fowl. It may be abbreviated to W. T. W. Jap. Bant. The other two varieties are in the same way made W. Jap. Bant., and B. Jap. Bant. The non-Standard Gray Japanese Bantam must be Gray Jap. Bant., for G. generally stands for Golden, and would be misleading.

Names of Silkies, Sultans, and Frizzles are not abbreviated, nor is it common to abbreviate names of either turkeys, ducks, or geese, though abbreviations of a few of them are admissible. On the contrary, it is more customary in some cases to use more words than is necessary. Thus a Pekin Duck is sufficiently described by that term, for there is only one variety, the White, but it is very common both in speaking and writing to say White Pekin Duck.

Most of the varieties of the turkey take their names from the color. There are Bronze, Buff, Slate, White, and Black. Then there is the Narragansett, which might be described as a bronze-gray in color. The White Turkey described in the Standard is called the White Holland Turkey. White Turkeys called Mammoth (a name also sometimes applied to the Bronze), have also been shown. Finally, there is the Bourbon Red Turkey which is perhaps best described as a Buff with the constituent colors unmixed, and tending to go each to certain sections instead of being evenly distributed.

The Pekin Duck takes its name from the port of Pekin, China. The Aylesbury, also a White Duck, from the district of England where it is most extensively grown. The "Standard" gives Rouen Ducks as of one variety, "Colored," but I do not remember ever having heard or seen the expression Colored Rouen Duck. The Cayuga Duck is Black, and it is quite usual to use the full descriptive title, Black Cayuga Duck. East India Ducks also are generally mentioned as Black East India Ducks. The Call Ducks, White and Gray, generally get their full titles. Muscovy Ducks, too, are fully described as Colored or White. The term "Colored," when applied to the name of a breed of poultry is, as is very evident, used as the readiest term to differentiate the variety to which it is applied from others which get a more specific description. Hence it may be observed that in the making as well as the using of names we may find evidences of easy going tendencies. A Colored Muscovy Duck is black and white. A Colored Dorking is a Dorking that is not white nor yet silver gray, but runs rather to shades of red or brown. A Colored Rouen Duck best deserves the term, for in the male, especially, there is a profusion of colors. The Crested White Duck is just a plain white duck with a top-knot. The Indian Runner Duck has a sort of hybrid name. It is not an Indian Duck, but according to best authorities, a Belgian production, and properly named and described as a simple Runner Duck, the name coming from its active habit. The Blue Swedish Duck takes its name from its color and from the country of its supposed origin.

Names of geese are seldom, if ever, abbreviated. The breed name may be used alone or with the color more specifically describing it. Toulouse and Embden Geese are quite frequently mentioned as Gray Toulouse and White Embden Geese, thus describing them more fully, though it is not necessary to do this to fix their identity. African Geese, described in the Standard as "Gray," are not at all such a gray as the Toulouse, but rather on the brown order. The Chinese Geese are the only breed having two varieties. These are Brown and White. It is quite common to call them China geese, instead of Chinese. The Wild Goose is also known as the Canada or Canadian Goose, and frequently given both titles, Wild Canada Goose.

LESSON XV.

Eggs and Egg Production.

IN this lesson we consider the egg and the subject of egg production from the "business" point of view, and for the common people not versed in any of the "ologies" into which the study of the egg is most fittingly introduced, or which may be applied most interestingly to the examination of the egg and the problems of egg production.

The egg owes its great commercial importance first of all to its food value. In it we have combined as in few other simple food articles the qualities of delicacy and substantial nourishing quality. Eggs alone constitute an important article of diet. But in the average family or hostelry their use as a separate article of diet is second in importance to their use as an ingredient in an almost endless variety of dishes. Except in the homes of the well to do the use of the egg as a separate article of diet depends largely on the relative prices of eggs and meat, the general tendency being to use eggs freely when they are cheaper than the popular cuts of meat, and to be economical in their use when the meats are cheaper.

The matter of cost also enters into the question of the free or economical use of eggs in cooking, though not to the same extent, for so many are the common dishes requiring eggs for their preparation that it is impossible for most housewives to make any considerable reduction in the quantity of eggs used in that way without entirely changing the ordinary bill of fare.

As lack of freshness and flavor are less noticeable in eggs when mixed with other articles, it becomes possible, and is customary to use as "cooking eggs" eggs which served separately to most people would be rather unpalatable. Indeed it is a matter of common knowledge among poultrymen that the infertile eggs from an incubator, tested out the fourth or fifth day, find ready sale to bakers — and possibly also to go into channels of more particular trade. Eggs preserved by various processes or kept in cold storage, and "held" eggs, that is, eggs kept without preservatives by the producers for weeks or even months in anticipation of rising prices, are also salable as "cooking" eggs. Of course these inferior eggs from various sources are not as good even for cooking as nice, fresh eggs, but so many people are satisfied to use them when the prices of fresh eggs are very high, that almost any kind of an egg that is not actually bad will sell readily for cooking purposes.

Eggs sell according to their quality as they reach the buyer. This is true as a general proposition — though some exceptions and seeming exceptions to the rule may be found. If they reach the buyer in as good condition as they left the producer there is no occasion for differences of opinion as to quality and value, but it is only when they go direct from producer to consumer that this is, generally speaking, possible. Most producers of eggs must send their goods to the consumer through channels of trade which require several transfers, more or less delay, and sometimes exposure to deteriorating influences. Sometimes the producer can exert some degree of control over the vicissitudes to which the egg in transit is subjected by investigating the course his goods take after leaving him, and selling to the middlemen who get them into consumption by the most direct route and with the least possible delay. It might be supposed that as a matter of business every dealer in produce would do that, but as a matter of fact there is a great deal of slackness in the handling of eggs, much more than in the handling of poultry, which more quickly shows deterioration whether alive or dead.

However, the producer's control of his eggs after their delivery to the first buyer or transportation company practically ceases as far as personal ability to protect their quality is concerned. Every producer of eggs can be very sure that there is no possibility of his eggs ever becoming better in quality than they were when produced. Age does not improve them to the normal taste, nor will any process of "ripening" render them more palatable. No tricks of manipulation will improve their appearance. The shipper of poultry may find the skill of the salesman who wipes, and shapes and makes more presentable the carcasses of the fowls, of benefit to him, making the fowls show often to better advantage than originally. But dirty eggs are graded as "dirties," and sold at an appropriate price. Small and misshapen eggs reduce the grade of their entire lot. Weak and watery eggs are readily detected by dealers and buyers. Bad flavored eggs in a line of good trade cannot come from the same place very many times in succession without someone in the line being called to account, and ultimately it comes back to the culprit among the producers.

When one begins to give special attention to the production of eggs, he must work for quality as well as for quantity, otherwise he gets but a part of the benefit of his efforts. Producing eggs in quantity and of good quality, he must market them to the best advantage. If he does not he may be no better off than he was in the first place.

Now let us take that the other way around. Suppose a man desirous of getting a better price for his eggs begins to study the ways of the markets, and finds that his eggs compete not with the best, but in the grade of cooking eggs. It is not impossible that there is discrimination or misrepresentation on the part of those handling the eggs, but it is far more likely that the eggs never were of the quality that they should be to command the best prices. This is especially the case with eggs from fowls for which most of the food is purchased, eggs from yarded fowls and eggs from fowls whose ration is too carefully balanced. It is a matter of common observation and frequent comment among eastern handlers of eggs that the western eggs as a class are superior in original quality to the eastern or nearby eggs. They are richer in color of yolk and in substance of white. At seasons when there is little deterioration in transit these western eggs may come into our eastern cities actually better than the nearby product, but during the greater part of the year time and exposure in transit operate to take away their freshness and flavor.

Now, as we have seen, it is easier for the producer to control original quality than to provide against a quick deterioration after the eggs leave his hands. Hence it should be apparent to the eastern producer that it is much easier for him to get good quality in his eggs than it is for the western producer to provide for the preservation of quality in his. There is really no excuse — but that of mistaken economy — for the producer near a good market not getting every advantage of price which excellence of product and nearness of markets combined should give him. Yet many producers do not get them. Why not?

Here are the principal reasons:

Debilitated stock.

Lack of variety in food and insufficient supplies of green foods and fats.

Excessive feeding of swill and other wastes.

Mistaken ideas of the food constituents required for egg production.

It takes healthy hens to produce eggs of first rate quality and fine appearance. Compare the eggs of individual hens in a flock, or better select certain eggs and then find and compare the hens that lay them. Your flock and their eggs may be too uniform in condition and appearance to make the comparison I suggest remarkable, but the average flock is not so. I succeed in having mine that way only in proportion as I limit my breeding to individuals carefully selected from stock bred in my own yards for generations. I find that when I go outside for new blood to improve some point in which I wish to make improvement, my most careful mixture of the new blood introduces a variety into the appearance and to some extent into the quality of the eggs which was absent during the years of close breeding. I find also that the lack of quality in eggs is coincident with a lack of vitality in the individuals producing them. It has sometimes happened, too, that in fertility my best layers and most vigorous birds were inferior to some of the others, and so I have sometimes had a larger proportion than desirable of laying hens from the weaker stock, and in such cases I have found the eggs averaging lower both in appearance and quality than when I succeeded in getting what pullets I wanted from my most vigorous birds. One can make more careful comparisons in matters like this in his own stock, but it is possible to see the facts in observations of the stocks of others. It is as unreasonable to expect first class quality in eggs from hens in poor condition as to expect good fruit from an unhealthy plant. A laying hen should be in good condition, with smooth, healthy looking skin and firm flesh. Some fat is desirable. Fat hens generally will lay richer eggs. A distinction should be made between fat and excessive fat, between healthy fat on an active fowl and the dead weight of fat an unhealthy fowl may carry.

What is variety? To different persons it means different things. A farmer may say that his cow gives so much milk or makes so much butter on grass. That may mean a considerable variety, though the one term *grass* covers it all. It is said that there are often as many as forty varieties of grass on an old pasture. Such a fact as this should be taken into account in considering the diet of hens on good grass range. On a western farm they may be fed nothing but corn, but they get also all the various kinds of grass which the pasture provides, many succulent weeds besides, and an almost endless variety of seeds of weeds and grasses, in addition to such waste grain other than corn as the farm may afford, and worms and bugs in great profusion. Compare such variety at this with the usual variety given hens kept in confinement, and it is easy to see where the greater variety is, and how meager by comparison is the variety afforded in a balanced ration containing even a dozen articles.

Where hens in confinement suffer most for lack of variety is in green and succulent food. Variety in grains is more readily provided. Grains are not perishable, and supplies can be kept on hand. But the dried substitutes for green foods, while excellent as far as they go, fall far short of the natural provision that way. Where fowls must be kept in confinement, and the ground room is very limited, I am inclined to think it is better to give up as much space as is necessary to the growing of vegetables especially for the fowls, even though by doing so the fowls are confined much more closely than desirable, and grow a variety of vegetables for them, lettuce, cabbage, rape, anything that they will eat.

The lack of fats in the ordinary ration results from overcaution in feeding fowls. To this is due the abhorrence of corn, which is far more prevalent in the east than it should be. A dealer in eggs in Boston who is also proprietor of a poultry farm has told me repeatedly that he had had many shippers whose eggs were so lacking in fats that they would not sell to the best trade, who had remedied the trouble by feeding corn. Indeed, he said, he always felt so sure that a shipper whose eggs were weak did not feed corn that he was in the habit of advising such to feed corn. Many handlers of eggs claim that they can readily distinguish between the eggs of corn fed hens and of hens fed wheat and oats and no corn by the appearance and consistency of the eggs when broken. Fats may be provided in other ways, but corn is cheapest.

The excessive feeding of swill is very common on "egg farms" near cities and towns where large quantities of swill and table waste can be had for the collecting, and there is generally a disposition to feed all of this that the hens can possibly be made to eat, and as little as possible of anything else. Weak and watery eggs and stock debilitated by an excess of soft food are common results. Much of the refuse food thus used is spoiled before being collected, and often the waste contains stuff the fowls ought not to have. Table waste properly saved and properly used is one of the best of foods for poultry, but feeding almost wholly on such food makes neither good poultry nor good eggs.

It is a common idea that for egg production "protein" especially is required. The fact is that what is needed in much larger proportion than it occurs in ordinary food articles is "fat." This fact explains why laying hens may be fed so freely of fattening foods and not only not become excessively fat, but even lose fat, sometimes. In this connection I would emphasize another point too generally unappreciated. The prevailing idea of egg production is that eggs are the product of such surplus of food taken into the body as a hen digests and assimilates, and does not require for the maintenance of other functions. This is a sort of half truth. Food taken in excess of current needs of the body for maintenance goes to eggs in one hen, to fat in another, causes digestive disorders in another. What makes the difference?

The attempt to answer that question brings us face to face with one of the most puzzling of the poultryman's problems — the control or regulation of egg production. Novices almost without exception suppose that expert poultrymen can regulate egg production. Experienced poultrymen know that when hens have started laying they can generally keep them laying, but that to assure the hens starting at or about any desired time is beyond their power.

Given a laying hen, and the volume of her egg production does depend very much upon the amount of food that she can use in excess of her bodily needs, though the maintenance requirements do not always take precedence. On the contrary it is quite a common thing for a laying hen's food to be diverted to egg production at the expense of bodily maintenance. When this continues for a long period the hen is greatly weakened, sometimes to the extent of becoming emaciated and exhausted beyond recovery. Such cases, however, are exceptional. The rule is that when egg production has appreciably exhausted a hen it ceases, and for a period longer or shorter according to the readiness with which the system is rebuilt all the energy of the fowl goes to restore it to perfect physical condition.

Generally speaking, it is correct to say that because a hen is laying she requires and takes food in excess of the needs of her body for maintenance and the performance of other functions, and that the volume of her product depends largely upon the amount of such surplus of food that she is capable of digesting and converting into eggs; but it is not correct to say that furnishing a surplus of food compels egg production and makes the hen lay.

What difference does it make which way we look at this matter? Just this difference:—Our way of looking at this matter is likely to govern our efforts to "make hens lay." If we believe that a surplus of the right kind of food will force egg production, we, very logically, devote ourselves to experiments with foods until we find one that seems to answer our purpose. If we believe that the activity of the hen's organs of reproduction depends upon something not so directly within our control as the kind, quality, or quantity of food furnished her, we are more ready to settle down to a good system, and have more patience in waiting for results when they do not come when we want them. It is conducive both to peace of mind and to continuing faith in a good method to know that egg production is measurably dependent upon causes or conditions beyond our control, and that failure to have hens begin laying when we want them to does not necessarily imply anything wrong — which by foresight or management we might have avoided.

LESSON XVI.

Some Elementary Moral Science For Exhibitors.

THE novice in exhibiting fowls finds, when he ventures into the show room, certain conflicts between the rules of exhibitions, as he reads them, and some common practices of exhibitors.

Thus the rules almost invariably say that specimens, except Games, which may have the combs dubbed, must be exhibited in their natural condition, but it is the universal practice among exhibitors to improve in various ways on the "natural" condition and appearance of the fowl. As to the legitimacy or illegitimacy of these practices, opinions may differ. All grades of opinion are found, from severe condemnation of even the most harmless and apparent forms of "grooming," to apology for or justification of practices which need only to be described to be recognized as wrong.

It is in the list of practices intermediate between these extremes that the things are found which cause the most concern to exhibitors who wish at the same time to be honest and fair, and to take advantage of every permissible method of improving a bird's chances of winning.

To the beginner in the exhibition room some of the practices which seem to the older exhibitor necessary and right appear to be of a very heinous nature. As he becomes more familiar with the conditions which occasion these practices, and learns to know the men he at first condemned for practicing them, he is very likely to considerably modify his views about both the men and the practices, even though he may not be able fully to approve them, or to join in them with a conscience entirely clear.

In this lesson I have no purpose or wish to persuade anyone to adopt a course with reference to these matters which he cannot justify to himself and his own conscience. I shall merely present certain facts in the relations in which they are commonly considered by experienced breeders and exhibitors, with the arguments by which the practices generally regarded as legitimate are justified, indicate the common attitude with regard to them, and occasionally give a personal opinion when it seems appropriate.

Some of the Evils of Exhibitions.

The great evils in poultry exhibitions as they discover themselves to the novice are:—

1.— The faking of birds; that is, treating them in some matter to circumvent a rule or gain an unfair advantage of a competitor.
2.— The borrowing of birds for exhibition.
3.— The collusion of exhibitors and judges.

What is Faking?

In the above classification of show room evils I have given a brief and comprehensive definition of a common term which means many different things to many different people.

Strictly and literally interpreted, the rules, as promulgated by most shows, do prohibit practices which it may be said are followed by experienced exhibitors without exception, for it would be utter folly for an exhibitor competing with experienced exhibitors in strong competition not to do these things. His chances of winning without them would be so rare that it would be foolish for him to enter his birds at all. Hence we may assume that a man who continues to successfully engage in strong competition in shows of any degree of importance does habitually do a number of things which the rules of the show say he shall not do. On this point there can be no dispute. The facts are self apparent to anyone who knows the conditions with which exhibitors have to deal.

Let us briefly examine these conditions:

To the visitor at a poultry exhibition, to the public at large, it is merely an exhibition, a display of fine fowls. For the exhibitors the poultry show consists of many competitions in the results of skill in breeding to a prescribed set of ideals, the full accomplishment of which, in combination, is practically impossible.

In some respects these standard requirements are absurd:—as when a fowl is disqualified for a defect inconspicuous until the fowl is subjected to a very close examination, or so obscure that its existence may not be positively identified without the aid of a magnifying glass. Were these requirements part of a consistent system they might be treated with more respect by exhibitors, but as the application of the same standards admits with trifling punishment and sometimes even without punishment blemishes and faults conspicuous as far as any quality of the fowl could be distinguished, the more familiar exhibitors become with the difficulties of producing fowls free from faults and with the incongruities in standard requirements the less evil they will see in disregarding or breaking rigid requirements about trifles.

Novices in the breeding of fowls rarely appreciate the scarcity of specimens which even approximately measure up to an educated conception of the requirements of the standards for their variety. The erroneous prevalent idea of the uniformity of thoroughbred birds and the fixity of characters in them is responsible for the common misapprehensions on these points as it is for many other difficulties of beginners.

To illustrate:—In many varieties what are known as "foul" feathers (that is, feathers not colored or marked as required) are likely to be found even in stock that has been most carefully bred. Indeed specimens on which an expert judge who made a thorough search for them could not find such feathers are very rare. The rule which requires specimens to be exhibited in their natural condition is commonly held to prohibit the removal of such feathers. It would generally be affirmed by officers of associations who might be questioned on that point that the intent of the rule was to prohibit the plucking of such feathers.

But the plucking of feathers which mar the appearance of the fowl, and the removal of which makes no visible defect or lack in the plumage is an act practically impossible of detection, after the job is done, and between this protection from consequences and the general feeling of exhibitors that the rule is unreasonable it has come about that exhibitors almost without exception — after a few seasons experience — pluck all the feathers that should be removed to make the bird appear at its best. Hence the rule is practically a dead letter except with novices who wish to strictly observe regulations and do not know the facts in regard to the common neglect of the rule, and do not appreciate the conditions which have made it obsolete.

While the facts given above do not justify a violation of such rules by those who consider such violation wrong, they do explain how it is that a great many exhibitors consider the violation of such rules an act involving no special moral turpitude. I have always maintained that such rules were wrong, because they could not possibly be enforced, and the disregard of unreasonable rules, while perhaps not of itself deserving severe condemnation, is to be deplored because of its effect on the observance of reasonable requirements. From this point of view I say that of the two evils the rule which prohibits fitting of this kind is the greater. The constant publication of such a rule also tends to confirm the prevailing error among beginners as to the possibility of producing fowls which are fit for exhibition without special attention to the removal of superficial faults. They naturally argue that if it were not a reasonable requirement it would not exist.

To show that such disregard of rules or laws is not peculiar to poultry exhibitors, and has been, and is, practiced by very large numbers of people without subjecting them to condemnation as particularly bad, let me cite the general disregard in this country of laws of the kind known as "Blue Laws," and the almost universal failure of people subject to taxation to return full schedules of their taxable property. If the reader disposed to be severe on poultry

THE GREATER EVILS. 135

exhibitors for their peculiar disregards of regulations will consider the number of regulations of all kinds which are practically obsolete so far as observance of them goes, it may dispose him to be more lenient in his judgment of them.

To me the evils of these practices, as between competitors, do not seem to be of as much importance as the evils which may follow after the exhibition. Every exhibitor understands —or has opportunity to know and understand — that such manipulations of birds as plucking foul feathers, washing white birds to give them the pearl white color required, coloring, or strengthening the color of legs and toes, etc., etc., are generally practiced, and knows also that the results of competition in the show room depend as much on the ability of exhibitors to condition their birds and fit them properly as upon their skill in breeding. The competition in the show room may, therefore, be considered one in which all meet on the same level, it being understood that each competitor, in addition to exerting his utmost skill to produce fine specimens, has also availed himself to the extent of his ability of the advantages to be gained by skillful preparation; it also being generally appreciated that birds which could go into competition without special fitting, with any hope of winning, are very rare.

Now, if with this general understanding the competitors meet, prizes are awarded, and each takes his birds home, we cannot readily discover that any particular harm has been done to anyone. The breeder who has exhibited birds whose faults have been treated knows what these faults were, and therefore can make an intelligent effort to eliminate them or reduce them in the progeny of the stock in the next generation. But suppose he sells them with the assurance to the customer that they have not been treated in any way, and the customer, relying upon his positive assurance, buys the birds, and perhaps in his ignorance of their fault mates them in just the way that will reproduce and perhaps exaggerate them in the offspring. Here we have a case which the consensus of opinion among poultrymen does not excuse, and while there are, of course, no data covering such points, I think it is a fact that in transactions between exhibitors the sellers will, with few exceptions, inform buyers of faults of this class. I have known many instances where, without stating their reason, exhibitors refused to sell such birds.

Another fact not generally understood by novices in exhibitions is that a fowl, having none of the blood of a variety, or perhaps but a fraction of blood of that variety, may to all outward appearances be a fine specimen of that variety, and the owner of such a fowl may exhibit it in the class to which, in appearance, it belongs without violating any rule of either the Standard or the association giving the show to which he sends the bird. The judge pronounces judgment on the specimens as they come before him. Neither he nor any officer of an association assumes to go back of the entries in considering the merits of a fowl. The Standard calls for certain peculiarities, but does not prescribe how they shall be produced.

I do not think it misrepresents the general attitude of poultrymen to say that they would see no special wrong doing in entering such a specimen for competition, but would consider it wrong for the owner of the bird to sell it except for just what it was. The distinction they make may not satisfy every requirement of a rigid moral code, but measuring the extent of an evil by its special results their attitude has something to commend it.

In a general way the extent to which a practice prevails affords a tolerably accurate measure of the degree of toleration or condemnation which the general opinion of the community or class interested assigns it, and the newcomer in a community and the novice in an interest alike need to be slow to condemn what may at first seem to them very serious and inexcusable faults commonly practiced by persons of general good character, for on further acquaintance with the facts it may develop that considerable justification of the practices in question may be found, and that toleration of them by those accustomed to them is as likely to be the toleration developed by a better understanding of conditions as the toleration of indifference to wrong through familiarity with it.

Buying and Borrowing Exhibition Birds.

Though I had cited only the borrowing of exhibition birds as an evil to be specially considered here, that can hardly be discussed as it should be without some reference to the buying of birds for exhibition.

Every novice in the show room, and I think I may say every veteran as well, is possessed of the ambition to put into an exhibition a string of birds of his own breeding that will win prizes enough to give him high honor as a breeder. To the novice it looks like a matter of buying good stock to start with and carefully selecting and reserving his choicest specimens. The veteran sees the matter differently. He knows that while it might be an easy matter to do that if he had the field to himself, with so many others striving to do the same thing there will almost invariably be a general division of prizes in any competition which is a competition in fact as well as in name. It is only at rare intervals that a breeder of a variety in which there is strong competition produces in his own yards as many first class specimens as he needs to enter in a strong class with reasonable expectation of getting his share of the prizes given. The really first class specimens are as a rule produced a few here and a few there — many of them by persons who either do not care to exhibit or would rather part with them at a good figure than take the trouble, risk, and uncertainty of winning in exhibition. Hence there are every year for sale a good many birds such as the breeders who wish to make large displays need to supplement their own production.

The rules of shows generally require that the bird exhibited shall be the *bona fide* property of the owner. That means that it must be his absolutely without any understanding or reservation. Occasionally at some show or in some special competition it is required that only birds *bred by* the exhibitor be entered by him, but as a rule the shows make no requirement of this nature going back of present ownership. The buying and selling of exhibition fowls cannot by any reasonable interpretation or application of common principles of right and wrong be made a wrong or even an objectionable practice. On the contrary, in its legitimate phases it may be said to be the most important feature of the interest in standard bred poultry.

But about this entirely legitimate feature of the business have grown up several abuses, most conspicuous of which is the lending and borrowing of specimens for exhibition.

This is carried on in two ways: — By simple borrowing and lending with not even a nominal change in the actual ownership of the bird; and by fictitious sale, or sale on such terms that it is substantially fictitious.

The practice began with simple borrowing and lending, but as poultry exhibitors generally frowned on it and general opinion would not condone it as it does some of the more prevalent forms of faking, those who wished to avail themselves of the use of exhibition specimens which they could not buy outright, and those who for various reasons were willing that their birds should be exhibited by others, devised the plan of selling birds conditionally, the bird to be returned after the show, and the price paid for it to be refunded. Such an arrangement is of course a mere juggle with right and wrong. The fiction of a sale does no more than make it impossible to prove the facts in the case until after the awards are made and the premiums paid. It does not often happen that birds "lent" in this way get back to their owner without interested competitors of the exhibitor finding it out sooner or later. Actual and positive proof of wrong doing and identification of birds is however so difficult that so far no effective check has been put on the practice. I do not think anyone has ever attempted to justify it. The advantages to be gained by it, both for those who borrow and those who lend, are so great that the temptation to make arrangements of this kind are very strong, and though the proportion of specimens in any show not actually the property of the exhibitor in whose name they are entered is probably always very small, I suppose that there are few exhibitors who have not at some time, perhaps in a very small way, yielded to the temptation to lend or to borrow. Many who would not exhibit birds not their own, have occasionally lent birds. Many who think the practice wrong have in emergencies borrowed birds. With the great majority such lapses have been exceptional, not habitual.

The disposition to lend — to do a fellow fancier a favor — is a manifestation of an excellent trait in human nature. With many fanciers the need of not indulging it does not become apparent until, having indulged it, they find that they must share in the common condemnation of the borrower.

The opportunity to borrow, say, at the time when a loss of or injury to a specimen upon which an exhibitor was relying has greatly diminished his prospects of making a good winning, presents itself as an evil of very small importance compared with the loss from which it may save him; but unless of feelings more callous than is usual among poultrymen, the borrower in the end resolves that he will never do it again.

On the whole, it may be said of this particular abuse that while it is prevalent enough to be a very disturbing influence, the general attitude of exhibitors toward it, and the fact that of late there has been a good deal of serious discussion of practical ways of preventing it, give reason to hope that it is an evil practice which is doomed to become less and less prevalent. Much might be said of the results of this practice, but the limits of this lesson make it necessary to pass over that phase of the subject with the remark that it develops consequences similar to those consequences of faking which especially concern the purchasers of stock, and develops them in larger proportions and more aggravated type.

Collusion of Exhibitors and Judges.

Poultry judges, as a class, are scrupulously careful and honest in placing their awards. Such a statement may surprise some who have at the same time the opposite idea about poultry judges, and think they have observed that I try to make no statements I cannot maintain. But I make this statement deliberately from a tolerably large acquaintance with poultry judges, extending back over many years, in which I have watched their work in the show room, seen some of their mistakes there, talked with them, and heard them talk with others about their errors, and learned also of their shortcomings outside of the show room.

There are few positions in life where it is harder for a man to avoid the appearance of evil than in the position of poultry judge, and few men who in that position are not constantly called upon to meet situations where they must decide off-hand matters for which anyone would, if possible, ask time for deliberation. This is true of points which must be passed upon in judging; equally true and of more importance on points that come up with reference to his relations to officials and exhibitors. To some extent it is true of any transaction of any kind in which he may engage. For instance, a judge, as a breeder, may sell exhibition fowls to parties upon whose exhibits he never expects to be called to pass, and in the course of his judging engagements may recognize those birds. There are very few judges who, in such a case, would not endeavor to place the awards honestly and fairly, yet probably every judge who is ever placed in such a position has realized the difficulty of feeling sure that he was acting without bias either for or against this stock, and has realized also that if it received awards he would be charged with having favored it — charged with collusion with the exhibitor; and that if it failed to get recognition the exhibitor might charge him, as a breeder, with selling as first class stock which he, as a judge, would not award a prize.

There are few, if any, judges who have managed to keep clear of every possible ground for suggestion or suspicion of collusion with exhibitors. Most judges, without giving chapter and verse, would probably admit in a general way that at times they had erred in their relations with officials and exhibitors, as well as made mistakes in the placing of awards. With so many exhibitions, so many judges, and so many exhibitors, there is always somewhere something that furnishes occasion for talk about the mistakes or the crookedness, or the vices of judges, and, given the occasion, there is generally a great deal more talk than the circumstances warrant. The result of it all is to give to many an impression of prevailing wrong doing by judges entirely out of proportion to the actual conditions.

While it is the little mistakes and little errors of judges that furnish most of the material upon which people build the opinion that crookedness prevails, there are unquestionably sometimes very wrong things done by judges, and often in such cases there is good reason to believe that they are done deliberately. Whether the judges who do them are indifferent to common standards of right and wrong, or feel so convinced of their own integrity and reputation that they think they may disregard appearances, I do not know. The general poultry public, and the novices who form an uncomplimentary opinion of judges, hear comparatively little of the larger and worse instances of delinquencies of judges.

Usually, with experience in poultry shows, a wider acquaintance with judges, and more particular knowledge of their faults, and of the judges who oftenest furnish occasion for criticism, one outgrows the attitude of general condemnation, and applies his disapproval more specifically, and with better discrimination. With a more correct appreciation of the situation I think an exhibitor is much more likely to so govern his conduct toward his exhibits and the judge as not to put the judge in an equivocal position. Exhibitors who do this become more careful on their own account as they learn that while they may, by attempting to act on the supposition that a judge's decision may be influenced, put him in an equivocal position, their acts are observed by other exhibitors, and these may feel doubtful about the judge in such cases, but their attitude toward the offending exhibitor is one of unequivocal condemnation.

An exhibitor who feels tempted — and perhaps especially tempted because he has imbibed the notion that judges are crooked — to tamper with the judge, may be interested in knowing that instances of judges being directly influenced in the placing of awards are extremely rare. Almost as rare are instances where a judge does not resent any palpable attempt to influence his decision. The common cases of apparent bias or prejudice for or against an exhibitor depend generally on the fact that the occasion takes the judge off his guard — that is, the judge's errors are unintentional — unconscious. Probably if anyone had the same interest in circulating stories of instances where the judge had shown a judicial cold blooded disinterestedness in the exhibits of his friend it would be found that such instances were many times more numerous than the others.

The general disposition of officials at shows and exhibitions is to hold judges to a very high standard of work, and in their dealings with those whose interests their decisions in any way affect. At the same time a correct appreciation of the conditions under which their work has to be done makes the veteran show official or exhibitor much more lenient in his judgment of a poultry judge's occasional shortcomings, and he is not so ready to utterly condemn a judge for things which while objectionable or deplorable do not seriously affect his efficiency or irretrievably ruin his reputation with discriminating fanciers.

LESSON XVII.

Business Morality in Poultry Culture.

WHOEVER becomes interested — ever so little — in thoroughbred poultry culture discovers almost at once that a considerable proportion of persons similarly interested believe that the moral tone of the industry is distinctly below the average. He will find many people who believe and say this, and many others who go further and declare that this extent of business immorality among poultrymen has so disgusted them with the business and those engaged in it, that they have either withdrawn from it entirely, or limit their active interest in it as much as is necessary to keep them quite strictly apart from those who indulge in or condone the practices which they condemn.

It is wise and well to take such statements with a liberal degree of allowance for the accuracy of the narrator's information as to general conditions and the correctness of his representation of his own case. The poultry industry, like every other, has its peculiar conditions offering temptation or inducement for peculiar manifestations of the errors of omission and commision to which human nature is prone, and the well known rule that men are much more impressed with the exceeding sinfulness of sin with which they are not familiar, applies here as elsewhere. So that it may readily be admitted that such evils as are complained of do exist, and that they do make a very strong impression upon the minds of many who see something of them.

Whether these evils are such and of such proportions as to give an uncommonly low general moral tone to poultry culture; and whether the persons who complain so much of them, and attribute their own lack of greater interest in poultry culture to them, are correct in their diagnosis of their case, are questions upon which I wish to make some comments before proceeding to discuss independently some of the real evils of the poultry business, their causes, and the means to be taken for minimizing them.

General Morality in the Poultry Business.

I think that, with very few exceptions — perhaps without exception — those familiar with the general conditions in the poultry business, and intimate with a great many men engaged in it, will agree that the general moral tone in the industry is the same as the general moral tone of the community. That means that, on the whole, the transactions of poultrymen and between poultrymen must be satisfactory to all parties concerned.

Now we know that it is possible in some kinds of business for those engaged in the business to do a dishonest business and still hold a large proportion of their clients or customers. This is accomplished by concealing the dishonesty of transactions, by deceiving customers as to their character. In the poultry industry by far the greater number of the acts of crookedness alleged to have occurred are of such character that to continually deceive the same persons by them is impossible. In fact, they are acts which — if actually committed — are detected even by a tyro in the business, with comparative ease. A bucket shop operator's victims cannot readily discover the mechanism, or follow the intricacies of the methods by which their separation from their money seems to be the result of causes beyond the control of the operator, but the man who buys a fowl that does not answer the description given him, and who finds that other people buying from the same parties receive the same treatment, has to do with a very plain and simple case of error or fraud involving nothing beyond the personalities and judgment of the two parties to the transaction. Hence it is theoretically true, (and, as a rule, is found to be true in practice), that it is impossible for poultry breeders either to perpetrate colossal frauds, or to continue indefinitely a regular system of dishonest dealing. So true is this that it is commonly said that a fraud is more easily detected and more quickly compelled to go out of business in this industry than in any other line. As a general proposition, I believe that this is true, though occasionally we find a case which we think is the exception — a breeder of and dealer in poultry commonly believed to be habitually dishonest, yet able to go on doing business, and apparently a good business, for a long period. Such a man's successful dishonesty may be explained by unusual capacity for getting the benefits of crookedness without incurring its penalties. Such "crooks" in the poultry business have their counterparts in every calling. In no legitimate calling do they establish the moral tone of that calling.

How Some Failures Let Themselves Down Easily.

One of the commonest things in everyday life is to hear men assign for their conduct in a matter, or for any condition for which they might be censured, a reason more creditable to themselves than the true reason. This is not always done with deliberate intent to deceive others. Often the person giving the reason deceives himself first. He looks for a reason that suits him, and, having found one, takes it as sufficient for himself. I would not say that everyone who gives as a reason for his failure to develop his interest in poultry the crookedness of the business was a failure in what he had tried to do in it. I would say, and I think the concensus of opinion of well informed poultrymen who have thought the matter over will agree with the statement — that most of the persons I have known who have given this reason for going out of poultry, or doing little with it, have been persons for whose failure, or lack of interest, experienced poultrymen who knew them would have assigned other reasons. Theirs is merely a case of "sour grapes."

Because of the frequency of instances of persons who not having realized their expectations in poultry culture attribute their lapses of activity to the evils of the calling in general, or to the deceptions or frauds of specified individuals or concerns, I advise those wishing to form for themselves a true estimate of the matter, to keep the point I have just mentioned in mind, and not to accept an explanation discrediting the calling generally from men who individually were no credit as fanciers or poultrymen either to themselves or to the fraternity.

Peculiar Conditions in the Poultry Business.

To properly appreciate moral conditions in the poultry industry it is necessary first of all to recognize in it certain peculiar conditions which foster what we may call the "sins of ignorance" — the mistakes of novices which furnish a much larger proportion of the transactions which might at first seem fraudulent than is commonly supposed.

In thoroughbred poultry we are dealing with a commodity in which good judgment of values cannot be acquired quickly, because the adjustment of values is a very complex problem. At the same time we are dealing with a commodity of the class in which, as a rule, novices who are much interested greatly overvalue their own judgment, because they do not realize how much values depend upon distinctions which, as novices, they are not yet able to make. To put it briefly and bluntly, the real cause of the failure of a great many sellers of poultry to do what they ought to do is ignorance of qualities and values in the goods in which they are dealing. This fact need not surprise anyone who will consider how common it is to see people beginning to sell thoroughbred poultry and eggs for hatching while their acquaintance with the breed or variety they handle goes no further than the stock they have in their own yards, and their experience with this may date back but a few months.

Now so far as the individuals in question are concerned, this period of ignorance of values is a stage in their poultry experience. Most of them outgrow it quickly as to serious errors, and quite completely within a few years. And so far as the industry at large is concerned, the presence in it of a class of novices who unintentionally make mistakes which are due to

INFLUENCE OF IDEAS ON CONDUCT.

ignorance, but may be attributed to crookedness, is a condition which, for all we can see now, must continue indefinitely, for each year the industry takes in a considerable body of new recruits, and there is no apparent diminution in the confidence of novices in their judgment of fowls. So we have always the same mistakes made, but made mostly by newcomers. I am speaking now of the great body of errors that furnish occasion for charges and rumors of crookedness and fraud, not of the occasional instances of intentional crookedness. Substantially all of those who continue in the business show by their later conduct in it that the errors of the first years were errors in judgment — not intentional frauds.

There is another side to the story of errors of this class. A considerable proportion of the buyers of poultry are even more ignorant of quality and value than the average novice selling poultry, and no whit less confident in their own judgment of these points. Such buyers are prone to find fault when no ground for fault finding exists; not always because they are disposed to find fault, but because their ideas of quality in fowls are badly distorted. These buyers, too, as they grow in experience and judgment, mostly pass out of the class who have many stories to tell to illustrate the prevalence of fraud and deceit in the poultry business.

While there are to be found here and there persons who, even after they are competent judges of the goods of different kinds handled in this industry and its allied branches, are still so much impressed by the evils they do see and meet that they see them out of proportion to the transactions which involve no crookedness, the bad reputation for morality, as is given the poultry business, does not in general get much confirmation from those who know the business. Among them it is regarded, as in fact it is, as neither better nor worse in a general way than other lines. The importance to the poultryman, whether a dealer or a buyer, of appreciating the real moral status of the industry is found in the connection between his ideas of business morality among poultrymen, and his own standards of practice in selling, and the attitude which he takes when buying. If one who has poultry to sell believes that the general moral tone in such transactions is low, he will often — perhaps unintentionally — be less careful in his own dealings than he would be if he believed that the usual practice was to give honest values. It is human nature to measure conduct by that of others, and to be satisfied if we can feel that we are a little better than the average. If one who is buying poultry believes that all dealers in fowls are rogues looking always for opportunities to defraud, and indifferent as to whether customers are suited or not so long as they get their money and escape the penalties of their practices, he is afraid to be satisfied with what he gets, and is apt to condemn it on general principles first, and then begin to look for specific faults.

Some Specific Alleged Evils of the Poultry Business.

When we say that the poultry business is neither better nor worse, on the whole, in moral tone, than the community, we admit that it contains a great deal of evil. I have already said that a very large proportion of the evils in the poultry business — of the wrongs done by poultrymen in their dealings with each other — consists of unintentional evils which most poultrymen avoid after they have learned wherein they were at fault. Another considerable proportion of the wrongs of which complaint is made consists of disappointments which come — according to my view — as phases of the ordinary risks of the business. In a great many cases these two classes of unsatisfactory incidents are mixed, both contributing to make the unsatisfactory situation. So, it would be difficult to make any hard and fast classification of evils according to causes, and I shall not attempt to do so, but simply mention a number of the most common sins attributed to poultrymen, and discuss each in order.

Doctoring Eggs for Hatching.

A beginner in poultry culture buys eggs for hatching, and gets nothing at all, or a few chicks from them. He is disappointed and sore. An acquaintance professing to be acquainted with the ways of poultrymen, suggests that probably the eggs were infertile, or were treated in some way to prevent their hatching. He will say positively that this is a common practice among breeders of fine stock who wish at the same time to get an income from their stock commensurate with its quality and reputation, and to prevent their customers becoming their successful competitors. So much is said and has been said with great positiveness in regard to the prevalence of this practice that a great many people who are by no means novices in the business believe that the reports of it must be true, and that it must be very general, or there would not be so many people so sure of it.

Considering the extent to which descriptions of evil constitute to some minds suggestions of evil, it would be strange if, with so much said of this evil, instances of it were not numerous, yet I have never personally known of a single instance of this being done; nor have I ever heard an instance of it reported by one whose testimony could be accepted as conclusive. Hence while I would not affirm that such an evil did not exist, I think that, considering the exceptional opportunities I have had of discovering it, that I am justified in asserting my belief that instances of such dishonesty are extremely rare.

The fact is, that there is no reason why any sensible or shrewd person should not want the eggs he sells to hatch, and there are many reasons why it is to his advantage to have them hatch well.

Substituting Eggs from Inferior Stock in Filling Orders.

This is an evil which, on their general impression of the unreliability of poultrymen, many suspect even when they have no positive proof of it. Absolute and positive proof of it is hard to get, but traces of it may be found on all sides. While it is probable that the great majority of breeders are strictly honest in this matter, I think that there is ample reason to believe that the number of those who, either as a regular practice or in emergencies, will substitute eggs that are not what the customer ordered, is very considerable. I base this opinion not on complaints of persons who think they have been imposed upon in this way, but upon the number of poultrymen I have found who, without actually admitting that they did this, would refer to it in such a way as to give the impression that they considered it a not very culpable offense; on a number of instances coming under my own observation when visiting poultry plants where it appeared that the orders for the best grades of eggs said to have been received could not have been filled with eggs from the stock of that grade; and on the occasional statements of men I believe to be trustworthy who told me that it was a regular or occasional practice on certain plants on which they had been employed.

With regard to the matter treated under the previous heading, I would say that the probabilities of a buyer of eggs being supplied with eggs which the seller had treated to prevent hatching, were extremely small, and the point need not be considered in ordering. With regard to the matter of filling orders with eggs not filling the specifications, I would consider that a buyer ran some risk, but not a very great risk. In what ratio this risk would be represented, I do not know. Probably one in ten would be an excessive estimate of the chances of getting an order made up in any part of eggs not as ordered, and I would suppose one in fifty a liberal estimate of the proportion of poultrymen who make a practice of giving eggs of grade inferior to those ordered. I offer these estimates only as indicating how much more rare this practice is than many suppose.

The application of general moral principles to the situation presents two phases:

First, there is the seller's side of the question. This phase of it presents no complexities. There is only one thing for an honest man to do, and one alternative: The one thing is fill orders with goods of the class and grade advertised for sale at the price. The alternative is to state his inability to fill the order, and to return the money.

The other phase of the question is not so simple. Many persons who wish to buy eggs for hatching must buy of poultrymen of whom they know nothing, or not buy at all. To say that if one cannot be sure of the honesty of the parties with whom he is dealing he had better let transactions of that class alone, is not to offer a practical solution of the difficulty. Moreover such a rule imposed on transactions in eggs is unreasonable because it puts on a transaction into which an unusual element of chance inevitably enters a rule more rigid than could be applied even in transactions from which chance might be almost completely eliminated. In other words, a person who takes the position that he will buy eggs or poultry only from breeders he thinks he is sure of, arbitrarily makes this a thing in which he will take no risks. It is his privilege to do that if he wishes, but doing it too often puts one in the list of those who are said to cut off the nose to spite the face. A more reasonable way to look at it is to make allowance for possible fraud of this kind, and consider it not as an outrage to be carried to the press, or into the courts, but as an ordinary risk, figuring the price of eggs which, for any reason give unsatisfactory results, into the total cost of the articles purchased.

To illustrate: Suppose A is a novice who is so situated that he cannot have any direct knowledge of any breeder of the variety of fowls in which he wishes to invest. Suppose he wants to begin with eggs. It is clearly impossible for him to learn of the different breeders' stocks and of their methods of dealing with customers in any other way than by buying of them. He must take some risks or not buy. The first man from whom he buys may not use him right. He may avoid a large loss by making his first investment small. If he is in a position to buy a considerable quantity of eggs he can reduce the risks of unsatisfactory results by dividing his order among several breeders. By doing this he is not likely to get as good results in the aggregate as he would have had if he had placed the entire order with the party or parties whose eggs gave him best results; but he knew nothing of what to expect when he placed the order, and we may assume that he has done much better on the whole than if he had placed the whole order with one of those whose eggs gave him poorest results. Further, while a single test and comparison of this kind does not furnish conclusive evidence as to the character of breeders and the quality of the various stocks, it does afford useful information on these points, and the beginner enters upon his next transaction with better assurance of getting what he wants.

One who will not buy until he is sure of those he deals with may easily lose more by waiting than another who takes chances will lose by the crookedness of some with whom he may deal. It is simply a case of "nothing venture, nothing have."

Boughten Eggs Which Hatch Unsatisfactory Stock.

The result indicated in the above heading is properly classed as an evil only when the unsatisfactory results are due to causes reasonably within the control of the person selling the eggs. It is not always possible to say in any particular case whether the seller or the buyer is responsible for unsatisfactory results. It is a fact, well known to poultry keepers whose experience has made them observe the point, that eggs from good stock may produce chicks which, under certain unfavorable conditions of care, feeding, and environment develop into specimens so inferior to the parent stock as to make it hard to believe that they are the progeny of that stock. The fact that the inferiority is due to such causes as are mentioned above is established in cases where from eggs produced at substantially the same time, chicks hatched by one party develop as would be expected, while those hatched by another party are a disappointment. Most of the inferior chicks from stock of fine quality are accounted for by lack of accommodations and lack of skill in those caring for them; but change of climate sometimes has decidedly unfavorable effects on chicks.

From this the reader will see that the fact that chicks from certain stock were unsatisfactory, does not prove that the breeder was dishonest. It should leave it an open question in the buyer's mind whether the fault lay with the seller, with himself, or with some person who had opportunity to change the eggs in transit. That is done to some extent, though how much it is not possible to say. It can be prevented by sealing packages. Some poultrymen selling eggs for hatching seal every package sent out, and advise customers to take notice whether seals are intact, and report if it is found that they have been tampered with.

The point we are now considering is not readily separated from the preceding matter in an effort to determine what is wrong in an unsatisfactory case. A breeder of poultry may send eggs that are not as represented, which yet give satisfaction in their results. He may send just what he advertised at the price, and the buyer get stock not at all up to his expectations. It may be a question then whether the breeder properly estimated the quality and value of what he offered for sale, or if the buyer is a competent judge of the value of what he produced. We cannot here follow the intricacies of such questions. I mention them to show the reader how impossible it is to make off hand or general decisions as to right and wrong in such matters.

We can, however, say that when a person advertising poultry sells eggs he would not expect to give him the results he knows his customers want, he is acting dishonestly, and that when the poultry keeper who is not quite sure of his own judgment of his stock, relies upon his judgment only in selecting and mating it for selling eggs for hatching, he makes a mistake.

As far as the buyer is concerned, he takes the usual risks on this as on other points when dealing with parties not known to him. There are times when, were he disposed to take the matter to law, he might secure redress in that way, but usually the amount involved is too small to make that worth while, and the common sense of most poultrymen leads them to charge such losses to experience, to avoid further dealings with those they find unsatisfactory, keep buying in sample lots wherever they think they are most likely to get what they want, and having found one or more breeders of the variety of their choice, whose stock and methods suit them, to do business mostly with those breeders.

Fowls That Are Not as Represented.

The number of fowls not up to descriptions sold each year is considerable when reckoned in numbers, yet not so impressive when compared with the whole number that changes hands.

Transactions in fowls are on quite a different basis from transactions in eggs. Generally speaking it is not possible for any marked change to take place in the appearance or condition of the fowl in the few hours, or, at most, few days, that intervene between its shipment by the seller and receipt by the customer. It may reasonably be assumed that instances in which fowls do not reach the buyer in approximately the condition they were in when packed for shipment are exceptional.

So if a fowl, on receipt, is found to be unsatisfactory we say that either the seller gave too little or the buyer expected too much, or that their ideas of what was wanted were so different that the transaction was on both sides a mistake. I have known many instances where people finding fault with the quality of the stock sent them had no occasion to find fault at all, the stock being just as represented, and the fault being in the buyer's ideas of what constituted quality. I have heard breeders vigorously denounced for having shipped a customer high priced stock decidedly inferior to some they had bought at bargain prices, when the conditions as to quality were just the reverse of what the buyer supposed, and the trouble was that his ideas were all wrong. When so much dissatisfaction of buyers is due to ignorance it is inevitable that there should be quite as many instances of people well satisfied with goods that are not worth what they pay for them. In this is found the breeder's greatest temptation to take chances in selling rather low grade fowls to people who do not appreciate quality, but want fowls that represent considerable sums of money.

The ethical and moral arguments that develop in considering this phase of the question are too deep for me. I have never tried to come to any definite general conclusions on them. I will here only briefly allude to a few of them that the reader may, perhaps, get some insight into the considerations which influence men with no wish to do wrong to do things which to many may seem very wrong. Let me give first, in illustration two points given to me by two very successful poultry salesmen, one mentioning one point, the other the other point.

A poultryman whom I was once visiting, discussing the matter of values and prices of fowls of different grades of quality, remarked that the controlling factor in fixing prices was not the actual or relative quality of the birds, but the number of people who wanted to own expensive fowls. In illustration of his point he told how one day a gentleman and lady drove to his farm in a fine turnout and wanted to look at poultry. They were much pleased with the birds in the first yard shown them, and asked the price of a trio. He mentioned a figure which probably correctly expressed the value of the fowls, say $25. Having named the price of these, he observed that the visitors lost interest in them. Being a shrewd man and experienced in the ways of buyers of thoroughbred poultry, he concluded that the price mentioned was too low. So he took them to a pen a little further along, and when they asked the price named a little higher figure; still further, and stopping before another pen, he priced what birds they wanted from that lot at seventy-five dollars, and a sale was quickly made. "You see," said he, "they didn't want fine fowls, they did not know or care anything about them. What they wanted was to have a few fowls that they could point out to their friends as having cost so much money. The birds I sold them at $75 were a little better than those I priced them at $25, but not much. But it would have been a crime to waste birds worth $75 on people who could not appreciate them, and only wanted to pay money for fowls."

Said another man to me one day:— "The art of selling thoroughbred fowls to make a good profit on them consists in finding out what a customer wants, and giving him a fowl that will answer his requirements. I have customers who want fowls I could not use, and would have difficulty in disposing of to others, and they are willing to pay as much for them as for birds I would consider good. Why should I take the position that only the points I and those who think as I do prize make quality? It is demand that makes prices. Breeders often have to breed to standards they do not like in order to sell their stock. I will sell a man anything I have that will suit him, and ask the highest price I think I can get. If I make a mistake, and the fowls do not suit him, I take them back and refund his money. If both the fowls and the price suit him, why is not that the value of the fowls? You may say he is satisfied because he is ignorant of standard requirements. That is none of my business. I cannot put myself in the position of assuming that a customer is ignorant; I have to take him at his own estimate of his knowledge of what he wants. How do I know but that he can make good use of birds worthless to me, and not salable to others? I hold that when an order is filled to the customer's satisfaction it is filled right."

In such bald statements or in the extreme instances of their application we see things which most of us unhesitatingly condemn as, at best, questionable. But the more we consider them the more we find that logically they lead us back to the question: "What makes the value of fancy poultry?" and we find it difficult to place a boundary line between what is and what is not permissible.

In practice the question of values seems to resolve itself into the question of suiting the customer, while most of the friction that arises between buyers and sellers is traceable to errors in what were really sincere efforts to please the customer.

Selling Unsexed Fowls.

Occasionally someone comes out and charges a breeder with having sold him "caponized" males or females. It is not likely that any breeder ever knowingly or intentionally did this. It is not improbable that breeders have often shipped fowls that were sexually impotent, and that in some cases post mortem examination would show a condition of the reproductive organs which was abnormal. Even in such a case it is not necessary to assume that the abnormal condition of the parts results from an operation, for it may result from disease of the organs which would not ordinarily be noticed; but in the case of a fowl in which the poultryman had special interest might be discovered because an effort would be made to determine what was wrong. It would be the height of folly for a breeder to castrate a fowl he intended to sell for breeding, or to sell a castrated fowl for that purpose. There is no authentic instance of it known.

Honesty the Best Policy.

While there are many points upon which people may differ as to honesty of certain transactions, I find no reason, as my acquaintance with poultrymen extends more and more, to change the opinion that, with rare exceptions, they intend to be honest. But if one had a leaning the other way, he would soon find that, as a matter of business policy, he could not afford to have many dissatisfied customers. Competition in the business is keen. It costs money in advertising to get a customer. It costs most breeders so much that if they had to depend on their new customers they would soon go out of business. After the breeder has once secured a customer he hopes to have him continue with him, and if he does not know he soon finds that while advertising may bring customers and continue to interest them, what holds the customer and brings further orders is satisfactory treatment.

LESSON XVIII.

Winter Egg Production.

MOST poultry keepers want to get eggs in the early winter when eggs are scarce and high in price. The difference between the fancier who says he does not care whether his hens lay them or not, and the poultry keeper who is greatly disappointed if they do not lay at that time, is not as great as at first thought it appears to be. Fanciers, as I find them, are not so indifferent to egg production as they sometimes profess to be. When their hens do lay well early in winter they are as pleased and as ready to boast of it as anyone. When they do not lay well at that time they console themselves with the thought that there are two strings to their bow, and that what they miss on early winter market eggs may be made up to them in the spring when they can sell the eggs for hatching. The poultryman whose eggs are not salable for hatching purposes has not another period of especially high prices to which to look forward, hence his disappointment over failure to get early winter eggs is greater, for he knows that his loss, if made up, must be made up from the profits of the remainder of the year on sales at lower prices. His need of winter eggs being greater, his desire to get them is greater; he plans for them and works for them, making it a point to have his stock ready to lay by winter if possible.

The fancier may be indifferent about the laying of such fowls as he intends to show, but for the rest of his stock he would, as a rule, rather have it laying than not laying. There are few fanciers who are indifferent to the receipts from market eggs, and fewer still who try to discourage egg production in any considerable part of their flock, for fowls eat nearly as much when not laying as when laying, and it takes but a small egg yield to pay the feed bills.

We may say then that the difference in the attitudes of practical poultrymen and fanciers in the matter of winter egg production is a difference in degree — not in kind — of interest. All want as many eggs in winter as they can get — but the intensity of desire, and of effort to get them, varies in a general way between these classes of poultrymen and also between individuals in either class. Perhaps the difference may be illustrated by a remark a friend of mine made to me one day at the New York show. We were talking of a man well known to poultrymen who has been a marked success as a money getter. Said he: "All men want money, but some will work harder to get it than others, and some will do things for money that others would not. Now the the difference between you and I and Blank in regard to money is this: If a dollar were rolling around on the floor, you and I would each make a grab at it as it passed us, but Blank would follow that dollar, on his hands and knees if necessary, until he got it."

In the ordinary course of events Blank will probably reach the age at which men retire from active life—if they can—with many times as much wealth as either my friend or I. He will get more because he cares more for it, and will work harder to get it. And this principle — or policy — (it is something of both) has a great deal more to do with the getting of eggs in winter than many would suppose. It has more to do with it than the kind of fowl, or the kind of food, or the kind of house. Within reasonable limitations it has as much to do with it as the period of hatching, the care and attention the chicks get while growing, and the treatment of the hens at the period when they are or should be laying. It is the intensity of the wish to have the hens laying in early winter, joined to a fair appreciation of the means of getting them ready to lay in winter, that makes some poultrymen work for this steadily and without intermission from season to season. I have never known anyone who was so uniformly successful in getting winter eggs that he or she might rightly be said to know how to "make hens lay," but I have known many poultry keepers who were much more successful than the average poultry keeper, and I think that those most successful in getting winter eggs divide quite naturally into two classes: Those who, in connection with more or less unnecessary "fussing" with the fowls, do the things essential to egg production; and those who do the essential things and nothing more. Poultry keepers of the first class are likely to get big egg yields and make large "per hen" profits; those of the second class to make more on their labor.

Essentials in Winter Egg Production.

One of the best poultrymen I know, a man who grew up in the business, and has been in it, on his own account since before he was out of his teens, once said that the only "secret" he knew anything about in getting winter eggs was to have the pullets ready to lay at the beginning of winter, and then give them enough to eat. The kind of food — within the ordinary range of poultry foods — he considered of little importance. Another good poultryman who violated many of the common rules of "correct" poultry keeping, speaking only of the handling of hens in laying condition, said that in his opinion the all important points were to give the hens an abundance of food, and to give it regularly. It was said by some poultrymen who knew him that they knew of no poultryman who could be away from home so much and yet get good results—better than any of those who looked after their stock much more closely. Asked about this, he replied that it was true that he took time off frequently in the middle of the day. He would not deny that at such times he "loafed," but he stated a fact which they had not observed, when he said that no one ever saw him leave or be away to the neglect of his stock, while various neighbors he mentioned would look carefully after their fowls for days or weeks at a time, and then for some reason or other there would be neglect, perhaps one feed omitted, perhaps the regular routine of feeding interrupted for one or several days, and this happening frequently, the hens did not get their full rations with the regularity essential to egg production.

Comparisons of the conditions, methods, and rations of poultry keepers and of the results they are getting, will clearly disprove any theory of breed, feed, or system of housing as superior to others, or as in itself essential to or assuring results, for in such comparisons we have to consider all kinds of results. We cannot consider only favorable results as advocates of special theories or ideas in any of these lines are wont to do. Comparisons such as I have just described would not so clearly prove the correctness of the ideas of abundant feeding and regularity as the essential things in egg production, for it is much easier in such cases to disprove a theory than to prove one, but such comparisons do generally indicate that good feeding —heavy feeding is essential to continued large egg production, and the reports and records of those who g t large egg yields in winter especially generally indicate that the fowls get regular care.

Conditions of Greatest Egg Production.

A fair comparison of results in different types of poultry houses will show — I think — the greatest winter egg production in the warmest houses, provided the ventilation of these houses is given proper attention. The same comparison would show very poor egg production, and often a great deal of sickness in flocks housed in this way. Investigation will generally show that in such cases the houses are not properly ventilated. The proper ventilation of warm tight houses by the doors and windows is a very simple matter—if the poultry keeper can open and close them as temperature conditions require. In theory this is easy — nothing could be easier. In general practice I have found that very few of those who use warm tight houses ventilate them properly. The common thing is to find houses still closed long after they should be open in the morning; then — if opened at all — left open long after they should be closed in the afternoon. The result of this is that the fowls become overheated and then chilled.

The proper way to ventilate such a house is to open the windows, or doors, or both, a little in the morning, wider toward noon; then close partially as the sun begins to get low, and altogether, or as nearly so as is customary, at dark. This for fair bright days. On stormy, cloudy, or windy days, the opening should be adapted to conditions. Here is where the tight house becomes a troublesome proposition to those who have to be away from the place much of the day, or find it inconvenient to vary the time of opening and closing doors and windows. The day starts bright, and everything is opened, the attendant leaving not to return until toward night. A storm of some kind comes up, and the house becomes very uncomfortable for fowls that are tender to such conditions. Or, the opposite case occurs. The morning is chilly and threatening, the houses are left closed. It clears, and the sun comes out warm, and the houses are overheated. It is such conditions that baffle the poultryman who with a warm house, hens fit to lay, and good food, has to take chances on the ventilation that is right in the morning being right through the day.

If one wishes to get the largest egg yield possible; if he can look after the ventilation properly; and if he is indifferent about the usefulness of the hens after the first winter, I would say, by all means use a warm tight house. One is surer of big egg yields in it. But if there are likely to be occasions when the ventilation of such a house would not be given proper attention, or if it is desirable that the hens should come through the winter in good physical condition, use a more open house, and be satisfied with the prospect of a more moderate egg yield. Bear in mind that the kind of house does not control the egg yield. It is only a factor — a factor which varies in value according to other conditions. The warm house seems to offer the greatest possibilities of heavy winter egg production, and at the same time to involve the greatest risks of poor egg production and debilitated fowls. It is a forcing house, and the dangers as well as the advantages of forcing are in it.

The matter of yard or range also seems to have an important bearing on egg production. The largest egg yields are almost invariably made by hens that are quite closely confined. The common experience is that as between two equal lots of fowls, fed as nearly alike as possible, but one confined quite closely and the other given a large yard or free range, the hens that are most restrained will give the better egg yield, often a very much better yield. The most reasonable explanation of this seems to be that the closer confined hens utilize all their food for maintenance and egg production, while the others put much of it into energy expended in running about. It is also reasonable to suppose that hens at large are more often frightened or disturbed, and it is well known that such experiences are likely to have a marked and immediate effect on egg production. Dairymen know that to get the largest possible flow of milk from their cows, they must be kept quiet and contented — not disturbed or frightened; but poultrymen do not so generally appreciate the effects of such things on the functions of the fowls.

Exercise is not always essential. By exercise here I mean compulsory exercise, compelling hens to exercise for much of the food they take. Undoubtedly many poultry keepers find that their hens do better when compelled to take exercise than when fed all they will eat, and taking almost no exercise; but a great deal of good laying is by hens which take little exercise. If hens have sound digestion and are not overfat to start with, they are likely to lay fully as well without much exercise during the early part of the winter, though as spring approaches they may get too fat or develop digestive troubles. We may say of cases where exercise is found necessary as (in a preceding lesson) of cases in which very careful attention to diet is found necessary, that these are abnormal. We may leave the matter of exercise in this way :— Exercise is not always essential; when it does appear to be essential provision for regular exercise should be made; it is always advisable if it is desired that the hens should go through the winter in good condition. If they are to be disposed of in the spring it does not make so much difference about exercise.

Generally the most convenient and satisfactory way of providing exercise it by littering the floors and feeding the grain in this litter.

Winter Rations for Fowls.

In this connection the reader should refer to the sample rations given in Lesson I., in the preceding series (1905). Indeed, it would be well to review the entire subject of feeding.

Those rations may be used as given, or, if it is desired to further cheapen the cost of feeding, the proportion of corn and corn meal in them may be increased, especially for hens that are evidently not overfat, or that are laying. After hens begin to lay it is not as necessary to guard against overfeeding and overfattening as it is with those that are not laying. When the reproductive organs are active the tendency is for them to take and use all available nourishment. When the reproductive organs are not acting the fowl, as a rule, eats less, though it may still eat more than is required for maintenance. When that is the case the surplus goes to fat. How far such fat as is accumulated prevents laying, is a question not yet satisfactorily answered. I think that there are relatively few cases where the ovaries of the hen are normal where any ordinary accumulation of fat prevents laying. There is some reason to suppose that the activity of the ovaries, and consequent production of eggs, are often retarded for months after the hen is otherwise fully developed, and that the reasons for this are not easily controlled. When this is the case a hen is likely to fatten, but when the ovaries do become active — which may be earlier in the winter, but is more likely to occur after midwinter — these fat hens and pullets usually lay a few abnormal eggs, and then lay normal eggs regularly — and usually such hens after beginning are heavy layers for that period. I speak of this because of the prevailing impression that slightly overfat hens will not lay — that there is a point in physical condition that must not be passed if hens are to produce eggs.

The conditions in winter admit of more latitude in liberality of feeding, as well as of the use of more of the "fattening" foods. Whole corn may be used quite freely during the cold weather, but as spring approaches should be fed with more caution, especially if the fowls generally show a tendency to become very fat, and they are to be kept through the spring and summer. Meat and bone may also be fed more freely than in warm weather.

For vegetable food clover, alfalfa, cabbage, mangels, and waste vegetables of nearly all kinds are used, and there is practically no danger of using too much of anything of this kind that is fed separately to fowls liberally provided with grain.

Rarity of Heavy Laying in November and December.

Novices in poultry keeping are quite generally under some misapprehension as to what is considered a good egg yield in these months. While occasionally better yields are obtained a yield of twenty to thirty per cent is an unusually good yield, and a poultryman who is getting as much as a ten per cent yield from his flock in November, has no reason to feel dissatisfied, and much reason to feel encouraged. Those who watch their flocks closely enough to get some idea of what individuals are doing, and of the relative proportions of pullets of the same age that are laying and not laying at this season are likely to discover that for most pullets the age at which they begin to lay is greater than the age usually given for laying maturity in their breed, and this knowledge can be turned to account next season by hatching enough earlier to have the bulk of the stock come to laying at the desired time, though the earliest layers may lay earlier than is desirable.

LESSON XIX.

First Treatment of Sick Fowls.

FOWLS are subject to a great many diseases. Quite all the more common diseases of men have their counterparts in poultry diseases. This fact is more generally recognized now than even a few years ago. It does not seem to be definitely established that diseases of like symptoms are identical in fowls, animals, and human beings. In the few cases in which we have reports of comparative studies of germ diseases of like nature in fowls and human beings, the conclusion reached has been that the germs were not the same. Some scientists, following the theory of development by evolution, suppose that as men, animals, and fowls were originally of one stock, so were the germs which produce certain diseases in them, but that many generations of life in a particular kind of organism, as a fowl, animal, or man, has especially adapted the germ to development in that organism and unfitted it for development in organisms of the other classes, and that while it is not impossible for a disease to be developed in any organism as a result of the introduction of the peculiar germ of another class of organisms, that result is extremely rare, and authorities are not at all agreed on the subject.

The practical value to the poultryman of a knowledge of this general fact of the similarity of human and poultry diseases is that it gives him a more reasonable attitude toward the diseases of poultry, and also enables him to apply such knowledge of the treatment of human beings presenting certain symptoms as he may have to the treatment of similar symptoms or conditions as they appear in his flock. There is no need of any poultry keeper, however inexperienced, standing in the presence of any of the poultry diseases most likely to occur in his flock helpless until he can get explicit directions from some poultry keeper or supposed expert on poultry diseases as to how to treat such cases, unless he is as inexperienced in the treatment of human ills as in those of poultry. Very few people who have arrived at an age, whatever that age may be in any case, when they can take the small responsibility of caring for a flock of fowls have not some knowledge of the treatment of the minor and more common human ailments, colds, indigestion, diarrhea, etc., for which there are many simple treatments in use. This knowledge can be applied to the treatment of ailing fowls, perhaps not always with the best results, but still as much better than doing nothing until specific directions can be obtained.

A sick person shows that he is not in good physical condition in a variety of ways, and first of all in extreme irritableness or in a reluctance to follow the every day routine of his life. The mature person whose sense of responsibilities impels him to continue his work when nature honestly rebels against it, is apt to be cross. With children unusual restlessness or unusual inactivity occur according to the nature of the trouble and the disposition of the child. With fowls, so much lower in grade of organization, and comparatively so deficient in brain and nerve force disease almost invariably means inactivity, separation as far as possible from the remainder of the flock, and a general attitude of listlessness or distress. Occasionally instances of the other manifestation of ill condition are seen, but they are rare.

Now the first step in the treatment of any trouble that has reached the stage where the fowl shows a desire for quiet and seclusion, is to furnish those conditions. It may be noted, further, that fowls—(I will not continue the comparison between fowls and humans, for as I proceed discussing the case for fowls, the reader whose attention has now been fixed on the

TREATING DISTEMPERS AND COLDS.

point will observe for himself how like human beings fowls are in these matters) — it will be noted that fowls in this condition try to find a comfortable place. The most comfortable place that affords quiet and seclusion, or as much of them as can be obtained in their quarters, is the place they take. Too often the poultry house and yard afford no suitable place for the fowl that is not fit to rough it with its companions. Especially is this the case when houses are stocked to their full capacity, and yards are small. Then it is often pitiful to see a sick chicken knocked about and run over by the rest of the flock, and thus deprived not only of the quiet it seeks, but of the strength it needs to concentrate on nature's effort to restore health. The natural tendency of the organization to recover its balance when weakened at any point, or in any function, is after all the most important factor in the treatment of poultry diseases, and he doctors best whose first step is to place the fowl in conditions where nature has a chance to begin the work of recuperation.

Take the ailing fowl away from the others, see that it has an opportunity to rest. Make it comfortable, remembering that what conditions will be comfortable for it depends somewhat on the nature of the trouble.

A fowl that is weak and debilitated by indigestion and diarrhea, and has a low fever, with symptoms of alternate chills and fever, will be most comfortable in a warm dry place. So will a fowl that seems to have poor circulation, that acts sluggishly, and the comb tends to turn dark.

A fowl that has a cold with collection of phlegm and mucus in the throat and nostrils, and discharges from the head, needs more than anything else, pure fresh air. It would be foolish, as it is unnecessary, to expose such a fowl to severe weather to give it fresh air, but it will generally be found that such fowls are benefited rather than injured by a degree of exposure much greater than most poultrymen think advisable for their poultry as a regular thing. Probably the best place for such patients is in an open coop in a sheltered spot.

Lame fowls should be put where they will be warm and dry. It is often difficult to determine the cause of lameness. Rheumatism is a frequent cause. With laying hens a strain when extruding the egg often results in a temporary or partial paralysis, which may disappear within a few hours, or, at most, a few days, if nature is given full opportunity to perform the work of recuperation. In all sorts of cases of lameness in hens in flocks in which there are males, it is especially necessary to remove the hen from the pen, for whether it is that the unusual attitude of the hen attracts his attention, or from some other cause, it frequently happens that a male forces his attentions on a sick hen in his flock to such an extent as to completely exhaust the strength of the hen. Even when the trouble is not so serious the recovery of ailing hens will always be more sure and rapid if it is impossible for the male to annoy them. For this reason it is advisable in cases of general indisposition in a flock, when all hens may be somewhat affected, though not enough to require isolation for all, to remove the male, when the hens will get along very well.

A point worth noting is the tendency, under certain climatic conditions, for fowls, animals, and people to have similar distempers in epidemic form. Conditions which result in many cases of a disease like "grip" or pneumonia among the people of a community are almost invariably accompanied by similar diseases in epidemic form among the fowls. Observing this, the poultryman will find it quite safe to treat the fowls for the same trouble for which the people are taking treatment. When medical treatment is to be given in such cases, give an ordinary full grown fowl the usual dose for a child of two or three years of age.

In applying external treatment for such troubles as colds people give hot foot baths, sometimes giving the whole body a hot bath or a sponging with hot water, then rubbing with lard or vaseline, or a mixture of such substances with turpentine or camphor, or both, or with a little carbolic acid. Such applications are very effective with fowls, though the mode of application must be varied. Instead of treating the feet and body of the fowl we steam and bathe the head, throat, and nostrils, then anoint with mixtures described.

When fowls are found with the face slightly puffed, or the eyes closed, and the lids gummed together try steaming with hot water and carbolic acid, (2 parts acid to 100 parts water) then rubbing with lard and carbolic acid in about the same proportions. Do this at intervals of two or three hours for a day, and in nearly every case taken in the early stages recovery will be immediate. Keep the fowl quiet and comfortable for a day or two before returning to the pen. Meantime try to determine whether there was any special reason in the conditions in that pen for a fowl to take cold that way, and if found, correct the trouble.

Fowls with diarrhea may be given a purgative if treatment is undertaken while the fowls, though somewhat distressed, are quite active, and eat and move about quite freely. But if the diarrhea has evidently greatly weakened the fowl it is better to check it promptly and to give a stimulant as well. For any of these purposes treat a fowl as you would a child of the age specified above, and you cannot go far wrong.

For a fowl that is crop bound, or that has the crop filled with gas and fluid as a result of disorders of the stomach, the first thing to do is to relieve the condition of the crop. It is in connection with such simple operations as these that the personality and deftness of the operator become factors. Some people are so rough in handling the patient, or bungle the operation so that the general condition of the fowl after relief is worse than before. Others will, without special instruction or experience, quickly and neatly do what is to be done. If one finds he is bungling such an operation badly he had better let it alone, and kill the fowl if the case seems too serious to be likely to recover without treatment.

Of what may be called minor ailments of poultry, I have found nothing so hard to deal with as the sweating and exhaustion that come from overcrowding young chickens in brooders or roosting coops. In these cases we have a combination of severe conditions continued for hours after the chicken has begun to be seriously affected by them. Either the crowding, or the overheating, or the partial smothering alone, if continued through the greater part of a night, would have very bad results. When the three are combined, and some chicks killed during the process, it is not strange that many of the survivors are so weak and exhausted that the system is very slow to begin to recuperate. How far very careful feeding, nursing, and care to see that each chicken was comfortable at night, would be successful in such cases, I do not know. With such attention as it is profitable to give ordinary chickens, those that have gone through an experience of this kind show the effects of it for months, or even all through their lives. This may not be observed if there is not another flock at hand with which to compare them, but when a poultryman's own stock furnishes opportunities for such comparisons he can hardly fail to notice it. My experience has been that, though some of them may turn out all right, on the whole it is more profitable to kill every chick in a lot that has suffered conspicuously from such conditions than to keep them for the sake of the few that may turn out well.

In cases of indisposition which might be due to food taken or some irritant substance taken with food, the best thing to do is to confine the fowls for a time where it is certain that they can get nothing but what the keeper gives them, and then feed only foods known to be pure and of good quality until the cause of the trouble has been discovered. Thus if a mixture of ground feed stuffs has been used, that is, an article sold in mixture, and there is any suspicion that something in it might be responsible for some trouble that has developed, discontinue its use, and feed only whole or cracked grains, or mashes made on the premises of known ingredients of good quality. In most cases of this kind careful feeding alone will soon bring the fowls back to health. I doubt whether it ever pays to do anything more than this for them. In all such cases one should do all that can be done to learn the cause of the trouble. If it is in the mixture of food used, discontinue that article, but be reasonably sure first that it is in that food. The trouble may be due to irregularity in feeding, and all that is necessary to do for it is to be regular in feeding and let nature work the cure. No change of food may be necessary —just regulation.

For mild attacks of indigestion occurring when there is no reason to suppose the trouble is due to the quality of food, there is nothing better — nothing so good, in fact,— as fasting, letting the fowls go without food, except green food, for one or two days then feeding sparingly. In highly fed hens the digestive system is often overworked. A system of feeding and quantity of food that one hen or one flock stands all right may be too much for another. The poultry keeper has to judge of what and how to feed, not by someone else's results, but by results in his own yards. To get good growth and heavy egg production we must feed heavily, and in feeding heavily we are always running risks of breaking down the digestive system by overwork. So the poultryman should watch closely for signs of indigestion, especially lack of appetite or looseness of the bowels, and when such symptoms appear let the hens go without grain for a feed or several feeds, as the case may seem to require. Careful attention to this point is the best preventive of digestive disorders. I have found it a good plan to omit one feed a week as a regular thing, and for many years have made it a practice to give the fowls one less feed on Sunday.

Another point of importance in the treatment of ailing fowls is to know when to let them alone — when to do nothing further than to put them in a quiet comfortable place, and let nature take its course. As an economical question I think that the poultry keeper who is wisest will adopt this sort of let alone treatment as his general policy, making exceptions only in the case of a fowl of unusual importance to him, or in cases where he feels sure a few very simple treatments will suffice. But apart from the economical question there is the question of when letting alone will be the best treatment for the fowl. This can only be determined by experience and experiment. When a number of fowls are sick at the same time, and with the same trouble, try treating a part and leaving the rest to recover if nature is able to work a recovery. You will be surprised to find how often the fowls that have no treatment recover just as quickly as the others.

Whenever a poultryman finds things going wrong in his flock, with no special cause for it that he can discover, he should sit down and consider whether there is anything in his situation or his methods, or any special condition existing, not in accordance with generally accepted ideas of correct conditions and methods, and whether any possible connection can be traced between his departure from usual things and the trouble that has arisen. In a majority of instances it will be found that the common practice of poultrymen is the safest to follow.

LESSON XX.

Poultrymen's Organizations.

IN concluding this series of lessons, and leaving for the present the method of treatment which in general has been pursued in the lessons of the past two years, I take the liberty of departing in a way from the method of selection of topics which I have heretofore tried to follow. The subjects treated in the forty lessons of the two series have been chosen because of the evident interest of poultrymen in them, and as far as circumstances permitted have been presented at times when they were most appropriate. The subject of this lesson is one in which too few poultrymen are interested, and most of those who are interested too little interested. The weakest point in the development of American poultry culture today is the lack of adequate organization of poultry interests and the almost universal indifference of poultrymen on the question of organization. True, conditions in this respect are improving somewhat, but unless the rate of improvement is greatly increased this generation of poultrymen will hardly begin to reap the advantages that must come to all when our poultry interests are efficiently organized.

We have in this country today these several kinds of organizations of poultrymen:—

1.—The American Poultry Association, a general organization in that it draws its membership from all parts of the United States and Canada, but so constituted that only a very few members can have any continuous activity in shaping its policies or directing its efforts. Poultrymen who are not fanciers rarely interest themselves at all in this association, and the average fancier is indifferent to it except when its existence is forced upon his attention by some inconvenience imposed upon him as a result of its manipulation of the standard descriptions of thoroughbred fowls.

2.—Various "state" poultry associations — a few of them properly so designated, but more of them merely local associations describing themselves as state associations either for the prestige the name may give them, or for the appropriation which may sometimes be secured from a state legislature for a state association, or one supposed to be of that class.

3.—A few sectional organizations, leagues of local associations; as yet none of these can be said to be completely organized and on an established working basis. So far they represent tendencies rather than actual accomplishments.

4.—Local poultry associations, organized primarily to hold a local poultry show and increase the local interest in fine fowls. Of these there must be now more than 400. The number is steadily increasing, and where a few years ago it was hard to find associations of this class that had been in active existence for more than two or three consecutive years, each season now adds substantially to the number which have had a continuous existence long enough to be regarded as permanent organizations.

POULTRYMEN NOT ENOUGH INTERESTED. 155

5.—Local poultry associations organized for instruction and information and exchange of ideas on poultry topics. Of these there are only a very few in existence. A few local associations organized primarily to hold shows also hold a number of meetings of an educational character during the year.

6.—Specialty clubs — organizations of breeders of a particular breed or variety; the object of the organization being to advance the interest of that breed or variety. These clubs generally draw membership from all parts of the country, though in a few instances where interest in the breed or variety to which the club was devoted was strong in some locality or section, the local members dominated the club. In such cases the government of the club is likely to be very democratic, the members generally attending the meetings and taking part in them, but the specialty club usually is an organization of which the secretary is during the period of his incumbency the autocrat.

Most readers will agree with me that the above enumeration shows variety enough in organization. As to the numerical strength of these organizations, it varies from less than ten to over a thousand.

But with all these organizations it happens almost invariably that when the occasion arises for organized effort on the part of poultrymen, either to advance or to protect their interests, there is no organization in the field so constituted that it can make its influence felt and its power respected by legislatures and corporations. The estimates of the value of poultry products which make them outrank many crops commonly considered as of greater importance are doubtless often exaggerated. But allowing a great deal for exaggeration, it still remains a fact that the value of our poultry products ranges well into hundreds of millions of dollars annually, and far exceeds the value of articles whose producers by combined and persistent effort are able to force a consideration of their interests on those responsible for state and national legislation. Why is it?

My answer is:—Because most poultrymen are too absorbed in the details of most intimate concern to them to have more than an occasional passing thought for the larger matters which should be of interest to all poultrymen; because poultry keeping being with most poultry keepers a side issue, the individual poultryman's financial dependence upon it is not great enough to strongly impel him to unite with others for the remedy of conditions that need improving; and because poultry keeping as a hobby, fad, or recreation draws its recruits largely from people of very modest means who have neither the money nor the inclination to make it as conspicuous, and their wishes or interests as much regarded by the rest of the people as some of the other forms of recreation. In fact, the average poultryman's disposition is quiet and retiring. The push and rush and noise of strenuous and spectacular sports do not draw him as do the quiet interest and occupation he finds in taking care of and developing his fowls. He may be and often is somewhat interested in popular sports, but rarely follows them with the zeal of their more pronounced devotees. A little of them will do for him; then back to the quiet, restful recreation he finds in poultry keeping.

Now, in poultry organizations, as in all associations, differences of opinion and of interest promote jealousies, disagreements, and divisions. The average poultryman would rather keep out of these, or drop out of the association in which they arise, than stay in and endeavor to work out a harmonious solution of the difficulty. The result is that dissension in a poultry organization usually results in its collapse, or, at least, in greatly weakening it, when if the members felt the importance to them of maintaining an organization dissensions would not so often be allowed to develop to the breaking point.

It may not be possible by presenting reasons for faults like this to persuade people to avoid them, but if such a statement of the causes of weakness in poultry organizations impresses the reader as true or reasonable it may serve to make his attitude in such matters more favorable to adjustment. To return to our subject proper:

These various poultry organizations all have claims on the attention of poultrymen when they solicit membership. As a general rule: Every poultryman ought to be associated with every organization which he can help, or which can be of help to him. But in the present condition of organizations many of us have to make exceptions to that rule, especially with regard to those organizations which draw membership from a wide territory, and still are controlled by a very few individuals.

In respect to what one ought to do in the present condition of poultry organizations, there is abundance of room for honest differences of opinion, but we find general unanimity in the view that every poultryman ought to belong to the local association in his town or district where such association exists. Even poultrymen who neglect to join their local association will usually admit that they think they ought to do so.

The local poultry association should be the unit of organization. No satisfactory general organization is likely to become established until poultrymen more generally appreciate the importance of maintaining a harmonious local organization, and the equal importance of alliance with other local organizations. Appreciation of these things is growing—but slowly.

The too common experience of local poultry associations is that within a year or two from a most auspicious beginning jealousies and disagreements reach the point where either the association divides, or a larger part of the members withdraw. For this the members individually are to blame, perhaps not all in equal measure, but it is rare to meet in such cases manifestations of the spirit of compromise which must exist in any organization which is to be permanent and efficient.

To this spirit every member can contribute. For the lack of it each member is responsible — for his share. There may be occasional cases where division or withdrawals are justifiable or inevitable, but to the impartial outside view such emergencies rarely exist, and in the great majority of cases, if the poultrymen who desire only harmony would work together in the interests of harmony, and not side with either one or the other of the opposing factions, but discipline both if necessary to bring about a satisfactory adjustment of the situation, its troubles would be short lived. And if, as a result of the common efforts of members of poultry associations who have no personal interest in the disputes which disrupt them, local poultry associations were made strong and permanent, as they should be; it would inevitably come about that these organizations would combine for the things that could be better accomplished or regulated through their combined efforts. While the tenure of life of the local association is as uncertain as it has been in times past, or even as it is today, that condition gives some warrant for the statement often made in defense of the autocratic and unpopular methods of some of the organizations of wider scope, that the lack of permanence in local organizations justifies the continuance of general organizations by methods constantly requiring the services of apologists and defenders. The uncertainty of continuance of local organizations also makes many of them reluctant to enter into league with others, and sometimes interferes seriously with the efforts to extend organization.

The individual poultryman who does interest himself in the matter, usually feels that he personally can do nothing to materially improve conditions affecting organizations of poultrymen, or that what he perhaps might do, he may not undertake because to do so would take time and attention which should be devoted to the occupation by which he makes his living. But what no one poultryman could do individually without a supreme effort, many individual poultrymen acting independently, though impelled by the same spirit, would accomplish easily, and if all poultrymen would interest themselves in their local associations, help those who are willing to bear the burden of the work of the association as long as their efforts are for the common good, and suppress those eager to take the lead when their efforts seem to be directed toward promoting their own interests or toward things of no value to the members, we would soon see far more healthy and vigorous life in local poultry associations, and a much more general intelligent interest in the constitutions and methods of the organizations of wider scope.

Beyond membership in local poultry associations I would not under present conditions insist on membership in any of them as a duty, for in them the individual member neither is, nor as they are at present, can be a factor as he may be in the local poultry association. In any of them a member may find or make opportunities to help along the cause of organization, but in the local association every man counts as nowhere else, and it is in the local associations that the foundations of a great and efficient poultry organization must be laid.

SOME THINGS THAT MIGHT BE DONE.

So in concluding this series of lessons I would like to say to all who have followed them, and especially to the many who have testified to the help they have found in them in such matters as feeding, building, breeding, etc., this question of an efficient organization of poultry interests is of as much importance to the poultryman — the one who stays a poultryman, as any subject in which poultrymen are interested.

In the course of these lessons I have at times imposed my opinions on readers. I have said of one thing and another: Take my word for this; or try this, and be convinced. As unhesitatingly I now urge on the reader who has not interested himself in organization, or has allowed his interest to lapse:— Interest yourself in and for a poultry organization in your locality. If there is one there already go into it and help and be helped. If there is none, organize one. Get together the few (there are always — nearly always — a few) interested in poultry, and have monthly meetings to exchange ideas. Thus you establish a nucleus about which local interest in poultry grows. Even the man or woman who is isolated in his interest in poultry has hope of companionship, for that interest is everywhere contagious.

Out of efficiency and permanence in local organization comes the power to protect poultry interests locally. This means a great deal in such a matter, as, for instance, a city ordinance prohibiting the keeping of fowls within the city limits. Usually a move to enact such an ordinance finds poultry keepers unorganized and unable to make any effective protest. Ordinances of this character are usually too rigid, arbitrary, and oppressive. There may be—there usually is — need of some regulation, but the entire prohibition of fowls within the city limits is not necessary, and if poultry keepers are organized and in position to make their rights respected they can secure such modification of a proposed ordinance as is desirable and fair.

As an illustration of what might be accomplished by concert of action by the local associations within a state, take the case of the proposed fox bounty law in Massachusetts a few years ago. This was a law to protect those engaged in an important industry. It was opposed by the fox hunters, who acted in concert, while the poultrymen did not. The result is that foxes flourish here increasingly.

As an illustration of what might be done by a powerful national organization, take the matter of express rates and regulations in regard to the transportation of fine fowls. Rates are unequal and often unfair, regulations mostly in the interest of the carrier, and to take away as far as possible protection to the shipper or purchaser. The express companies make and unmake rates and rules at will. Poultrymen find fault, protest ineffectively, and submit, because there is no collective force behind their protests.

These are but a few of the things that go wrong for lack of strong organizations of poultrymen, most of which would be righted with comparative ease if taken in hand by an organization having the united support of those whose interests were affected.

These matters may not obtrude themselves constantly on the poultryman as do those which we call the practical details of the business, but they are none the less vital. They can be dealt with only by concerted action, but concerted action can come only as individual poultrymen unite to make it. The responsibility is on everyone.

INDEX.

Abbreviations, 125.
Abdomen, development of, 118.
Accommodations for turkeys, 44.
African geese, 35.
American class, 88.
American Poultry Association, 154.
Artificial hatching of goslings, 39.
Art in poultry culture, 60.
Asiatic class, 94.
Asiatic fowls, 88.

Back, shape of, 118.
Barred Plymouth Rocks, 88, 91.
Bedding for ducks, 28.
Black turkeys, 44.
Body, shape of, 117.
Borrowing birds, 135.
Boughten eggs, 143.
Brahmas, 88.
Breeding geese, care of, 38.
Breeds, how many, 96.
Breeds of geese, 35.
Bronze turkeys, 43.
Brooding ducklings, 29.
Brown Leghorns, 93.
Buff Leghorns, 93.
Buff Orpingtons, 93.
Buff Plymouth Rocks, 91.
Buff turkeys, 44.
Building for ducks, 25.
Business ability, 56.
Buying exhibition birds, 135.
Buying stock, 98.

Care of breeding geese, 38.
Carriage, 118.
Catching turkeys, 51.
Chance in poultry keeping, 104.
Chart, Felch's breeding, 11.
China geese, 35.
Chloro-Naptholeum, 107.
Cleanliness in duck culture, 32.
Climate, effect of change of, 19.
Cochins, 88.
Colds, 151.
Color, 122.
Combs, shapes of 119.
Condition natural, 133.
Continuous house, location of, 84.
Cooked food for ducks, 28.

Cost of geese, 37.
Crangle's rations for turkeys, 48.
Cross bred fowls, 86.
Curtiss Bros.' rations for ducks, 28, 31.
Curtiss' ration for turkeys, 48.
Cushman's opinion of breeds of geese, 35.
Cushman's rations for turkeys, 49.

Dalton's law, 8.
Darwinian theory of inbreeding, 6.
Debilitated fowls, 151.
Diarrhea, 152.
"Doctoring" eggs for hatching, 141.
Dominant characters, 20.
Dorkings, 94.
Dressing ducks, 33.
Dressing turkeys, 51.
Duck culture, 24.
Ducks, growing for stock, 33.

Egg production, 129, 146.
Eggs, beginning with, 97.
Embden geese, 35.
Essentials in winter egg production, 147.
Exhibitions, some evils of, 133.
External characters, value of, 114.
External parasites, 106.

Failures in poultry keeping, 52.
Faking, 133.
Fancy points, value of, 115.
Farmers' poultry, 61.
Farmer, what poultry culture offers the, 70.
Farm for poultry, 77.
Fats in food, 131.
Fattening turkeys, 49.
Feeding breeding ducks, 28.
Feeding ducklings, 30.
Feeding goslings, 41.
Feeding turkeys, 48.
Felch's breeding chart, 11.
Fences for ducks, 25.
Fences for geese, 38.
Fickleness in poultry culture, 59.
Flavor in eggs, 129.
Flock, number of ducks in, 26.
Foul feathers, 134.
Fowls described, 86.
Fowls not as represented, 144.
French class, 95.
Fresh eggs, 129.

INDEX.

Game fowls, 95.
Gape worm, 111.
General purpose breeds, 89.
Golden Wyandottes, 92.
Goose culture, 34.
Grade fowls, 86.
Grass land for poultry, 75.
Green food for ducks, 28.
Green food for goslings, 41.
Green foods, substitutes for, 131.
Grove Hill poultry yards, 76.

Hallock's rations for ducks, 31.
Hamburgs, 88.
Hargraves' ration for turkeys, 49
Hatching ducklings, 29.
Hatching goslings, 39.
Hatching turkeys, 47.
Head characters, 119.
Heredity, 8, 17.
Honesty the best policy, 145.
Houdans, 95.
Houses for geese, 38.
How many breeds, 96.

Importations, early, 87.
Inbreeding, 5.
Indian Game, 95.
Indigestion, 152.
Indiscriminate breeding, 7.
Internal parasites, 110.
Intestinal worms, 112.

Judges, collusion of exhibitors and, 137.

Killing ducks, 33.
Kinds of fowls, 86.

Labor, 58.
Lame fowls, 151.
Laying out poultry plants, 75.
Learning poultry keeping, 59, 103.
Leghorns, 93.
Lice, 106.
"Like begets like," 7.
Line breeding, 5, 10.
Local poultry associations, 154.
Locating poultry plants, 75.
Location, 57.
Location for duck growing, 25.

Mackey's ration for turkeys, 48.
Market duck culture, 24.
Marketing ducks, 32.
Marketing geese, 42.
Mating geese, 38.
McFetridge's ration for ducks, 28.
Mediterranean class, 93.
Mendel law, 19.
Minorcas, 94.
Mites, 107.
Model plants, 79.
Mongrel fowls, 86.
Mongrel goose, 35.
Morality in poultry business, 139.
Moral science for exhibitors, 133.
Mouse, application of Mendel law to, 21.
Narragansett turkeys, 44.

Natural condition, 133.
Nest for turkey, 46.
Newman's rations for geese, 39.
Nomenclature, poultry, 125.
Novices' errors, 53.

Objections to inbreeding 6.
Obsolete types and characters, 19.
Ordering stock, 99.
Organizations of poultrymen 154.
Original ideas, 57.
Orpingtons 90.

Parasites, external, 106.
Parasites, internal, 110.
Partridge Plymouth Rocks, 92.
"Peaked" fowls, 120.
Pekin ducks, 24.
Phenomena of heredity, 18.
Plumage, 122.
Plymouth Rocks, 88.
Polish, 88.
Pollard's comparisons of geese, 37.
Pollard's rations for ducks, 28, 31.
Pomegranate bark for worms, 113.
Prepotency, 16.
Prepotent characters, 22
Profit, doing work at a, 54.
Pure bred fowls, 86.

Quality of stock, 97.

Rankin's ration for ducks, 28, 31.
Rations for geese, 39, 41.
Rations for turkeys, 48.
Rations, winter, 148.
Recessive characters, 20.
Recreation in poultry keeping, 74.
Red Caps, 94.
Returning stock, 99
Reversion, 18.
Rhode Island mongrel goose, 35
Rhode Island Reds, 90, 92.
Roost for turkeys, 50.
Round worms, 112.
Rudd's rations for geese, 39.

Salaries of poultrymen, 73.
Santonine for worms, 113.
Scalded food for ducks, 28.
Scientific feeding, 60.
Sectional organizations, 154.
Selecting breeding ducks, 27.
Selection, 8.
Selling unsexed fowls, 145.
Shape, 117.
Sick fowls, treating, 150.
Silver Laced Wyandottes, 92.
Silver Penciled Plymouth Rocks, 92.
Silver Penciled Wyandottes, 92.
Size, 116.
Skill, how acquired, 54.
Skin, texture of, 121.
Slate turkeys, 44.
Spanish, 88.
Standard, 86.
Starving for indigestion, 152.
"State" poultry associations, 154.

Stocking the poultry plant, 96.
Substituting eggs, 142.
Systematizing work, 58.

Tail, development of, 120.
Tape worms, 112.
Temperature of duck brooder, 30.
Testing stock, 101.
Theoretic line breeding, 11.
Thompson & Co.'s plant, 81.
Toulouse geese, 35.
Turkey growing, 43.
Two thousand hen plant, 81.
Typical ventures in poultry keeping, 61.

Unsexed fowls, selling, 145.

Values of external characters, 114.
Variations, 18.
Varieties of turkeys, 43.
Varieties, one or more, 96.

Variety in food, 131.
Visiting poultry farms, 85.
Vitality, loss of, 7.

Wages of poultrymen, 73.
Washing fowls, 135.
Water for ducks, 32.
Weber Bros.' rations for ducks, 28, 31.
Weights of geese, 35.
White Holland turkeys, 44.
White Leghorns, 93.
White Plymouth Rocks, 91.
Whitewash, 107.
White Wyandottes, 92.
Wilbur's rations for geese, 39.
Wings, size and shape of, 120.
Winter eggs, 147.
Winter rations, 148.
Worms, 111.
Wyandottes, 89.